www.wadsworth.com

wadsworth.com is the World Wide Web site for Wadsworth and is your direct source to dozens of online resources.

At *wadsworth.com* you can find out about supplements, demonstration software, and student resources. You can also send email to many of our authors and preview new publications and exciting new technologies.

wadsworth.com
Changing the way the world learns®

Generalist Case Management

Generalist Case Management

A Method of
Human Service
Delivery

Marianne Woodside
Tricia McClam
University of Tennessee, Knoxville

THOMSON
BROOKS/COLE

Australia • Canada • Mexico • Singapore • Spain
United Kingdom • United States

Sponsoring Editor: *Julie Martinez*
Marketing Manager: *Caroline Concilla*
Marketing Assistant: *Mary Ho*
Advertising Project Manager: *Tami Strong*
Assistant Editor: *Shelley Gesicki*
Editorial Assistant: *Mike Taylor*
Project Manager, Editorial Production:
  Kim Svetich-Will
Production Service: *Publishers' Design and
  Production Services, Inc.*

Copy Editor: *Joan Kocsis*
Permissions Editor: *Sue Ewing*
Cover Designer: *Denise Davidson*
Cover Image: *David Ridley*
Print/Media Buyer: *Vena Dyer*
Compositor: *Publishers' Design and
  Production Services, Inc.*
Printing and Binding: *Transcontinental
  Printing Inc.–Louiseville*

For more information about our products, contact us at:
Thomson Learning Academic Resource Center
1-800-423-0563

For permission to use material from this text,
contact us by: **Phone:** 1-800-730-2214
Fax: 1-800-730-2215
Web: http://www.thomsonrights.com

Library of Congress Control Number:
2002107344

ISBN 0-534-54897-0

Brooks/Cole—Thomson Learning
511 Forest Lodge Road
Pacific Grove, CA 93950
USA

Asia
Thomson Learning
5 Shenton Way #01-01
UIC Building
Singapore 068808

Australia
Nelson Thomson Learning
102 Dodds Street
South Melbourne, Victoria 3205
Australia

Canada
Nelson Thomson Learning
1120 Birchmount Road
Toronto, Ontario M1K 5G4
Canada

Europe/Middle East/Africa
Thomson Learning
High Holborn House
50/51 Bedford Row
London WC1R 4LR
United Kingdom

Latin America
Thomson Learning
Seneca, 53
Colonia Polanco
11560 Mexico D.F.
Mexico

Spain
Paraninfo Thomson Learning
Calle/Magallanes, 25
28015 Madrid, Spain

About the Authors

Together, Tricia McClam and Marianne Woodside have almost 50 years of experience in human service education, as well as many years working as practitioners in education and vocational rehabilitation. Currently they are professors in the Human Service Education Program in the College of Education at the University of Tennessee. They are committed to research in teaching and learning in the human services and have written two other texts, *An Introduction to Human Services* and *Problem Solving in the Helping Professions*.

Contents

3   Models of Case Management   61

7 Building a Case File   163

8 Service Coordination   211

9 Working Within the Organizational Context   239

10   Ethical and Legal Issues   267

• • • • • • • • •

11   Surviving as a Case Manager   291

• • • • • • • • •

Preface

For us, the purpose of writing textbooks is to share with our students and colleagues what we have learned about the profession of human services during our years of teaching and working in the field. The preparation of this second edition of *Generalist Case Management* has been guided by this philosophy. Primary informants for this edition were both educators and human service professionals. Through our associations with colleagues in professional organizations and educators who used the first edition, we have learned about current trends, challenges, and new knowledge and skills necessary for effective case management. Interviews we conducted with case managers across the United States since 1994 enabled us to capture their voices as they described the realities of service delivery. We believe this adds a real-world perspective to the text.

Change occurs rapidly these days, and case management as a service delivery strategy is no exception. Factors affecting case management today include federal legislation, emerging client groups, technology, shifting demographics, new service delivery models, and the resulting ethical and legal dilemmas. The second edition of *Generalist Case Management* reflects these changes. There are updated references and examples, a focus on strengths-based case management, a review of technological advances, and an emphasis on collaboration that embraces family, friends, and clients as case managers. New trends and challenges in case management are identified and discussed. Finally, a new emphasis on diversity in its broadest sense—ethnic, religious, gender, and lifestyle—pervades the text.

The concept of case management is dynamic. Just as the process has changed during the last decade, so it will continue to evolve during the 21st century. Many factors will influence human service delivery in the future: economic instability, the managed care environment, technology, the scarcity of resources, demands for accountability, and the changing political climate. We have defined and described case management as it is practiced today, but with an eye to the future.

OUR GOALS AND OBJECTIVES

We have written this book to serve as an introduction to the concept of case management and explain how it is used to provide human services. In addition, we expect this text to serve as a reference as you begin your fieldwork and then your professional career. Our in-depth approach to the study of the case management process will make it valuable for you to review relevant chapters when confronted with a challenge or a dilemma in your work.

We begin by defining case management, discussing its history, and describing the models that are used. The case management process is traced from the intake interview to termination. Professional issues and skills are explored, and the most up-to-date aspects of case management are discussed. In short, our goals for this text are fourfold: to define case management, to describe many of the responsibilities that case managers assume, to discuss and illustrate the many skills that case managers need, and to describe the context in which case management occurs. Underlying these goals are the human service values and principles that guide them.

ORGANIZATION OF THE TEXT

The first three chapters of *Generalist Case Management: A Method of Human Service Delivery* focus on defining case management. Chapter 1 introduces the concept of case management and describes the context in which human service delivery occurs today. This chapter begins the exploration of the differences between traditional case management and what is practiced today. Roy Roger Johnson's case illustrates the three phases of case management: assessment, planning, and implementation. Chapter 2 expands the definition of case management by reviewing its history. The case of Sam, who was institutionalized early in childhood, illustrates how the changing definition of case management has been reflected in the care of clients. First-person accounts of clients in the early days, as well as excerpts from relevant legislation, enliven the history. Managed care, which has a strong influence on human service delivery today, is defined and discussed in terms of its effects on the case management process. Chapter 3 introduces the roles and responsibilities assumed by the case manager and the models used in service delivery. Vignettes and examples illustrate the many roles of the case manager and the various models of the process.

Chapters 4 through 8 describe in detail the phases of the case management process. Chapter 4 introduces the assessment phase, emphasizing the application for services, case assignment, and documentation. Conducting an intake interview, evaluating an application for services, and reviewing the collected information occur during this phase. Guidelines for documentation conclude the chapter. Chapter 5 focuses on the intake interview, including attitudes, characteristics, and skills of interviewers and pitfalls to avoid while interviewing. Chapter 6 introduces planning, the second phase of case management. The activities and skills needed to work with clients and colleagues are the focus of this chapter, which includes such useful topics as developing a plan of services, identifying service providers, and gathering additional information. Tests and their appropriate uses are discussed. Components of the case file are the focus of Chapter 7. Medical, psychological, social, vocational, and educational information complete a picture of the client. Chapter 8 focuses on the case manager's interaction with other colleagues. A discussion of service coordination explores the process, including referrals and effective communication with other professionals. Advocacy, a major responsibility of the case manager, is discussed in depth. This chapter also examines how to work effectively as a team member and as the team leader.

Chapters 9, 10, and 11 address the context in which the case management process occurs and introduce the professional issues case managers encounter. Chapter 9 covers the organizational structure of human service agencies. Chapter 10 presents legal and ethical issues confronting the case manager, including confidentiality and autonomy. Other relevant issues are working with violent clients, the duty to warn, and the question of when to break the rules. Chapter 11 concludes the text by introducing the knowledge and skills that case managers see as critical. Strategies to survive as a case manager include avoiding burnout, using time management, and asserting oneself.

ACKNOWLEDGMENTS

Many people contribute to an undertaking such as this text, and we would be remiss if we failed to acknowledge them. Our colleagues in the National Organization for Human Service and the Council for Standards in Human Service Education have encouraged and supported our efforts to investigate case management—offering suggestions, reviewing materials, and sending information. Our colleagues Steve McCallum and Amy Skinner provided valuable information about assessment. Our students continue to be a major source of experience, assistance, and feedback—particularly Kylie Cole, our graduate assistant. Ray Vaughn also has our gratitude for sharing his perspectives on case management.

The case managers that we interviewed during the past five years made many contributions to this book. They shared their time, experiences, successes, and failures to enlighten us about the complexities of case management. It is their words that give this text a firm grounding in reality. Among their contributions are definitions of case management, perspectives on the components of the process, and evidence of the trends and challenges that the future holds. Most of all, we thank them for helping us understand the dynamics of the rich and varied process of case management.

Throughout our careers we have valued the review process. The comments and suggestions of Bonnie Bedics, University of West Florida; Mary Ann Bromley, Rhode Island College; Kan V. Chandras, Fort Valley State University; D. Shane Koch, University of North Texas; Ken Korpi, Dawson College; Robert Martin, Skagit Valley College; Carol Murray, Kent State University–Ashtabula; Sheri Narin, Piedmont Community College; and Barbara Peterson, Tacoma Community College, were critical to the development of this text. As they read the printed version, we hope they will be able to see how their unique contributions have improved the text.

Of course, our friends at Brooks/Cole deserve our thanks. Their expertise and assistance have been central to the project. They include Julie Martinez, editor; Shelley Gesicki, assistant editor; Kim Svetich-Will, project manager; and Sue Ewing, permissions editor.

Last, but not least, we thank our families for their support during this effort. We have spouses who encourage our writing and support us in our academic endeavors.

As the field of human services continues to grow and develop, we look forward to hearing from you. We hope you will share with us your observations and experiences with case management in the field, as well as your reactions to this text. Please send us your comments.

Marianne Woodside
Tricia McClam

Introduction to Case Management

*T*here is only one mission here and that is to see to it that persons with very complex medical needs go home and stay home safely. This agency delivers services to individuals with over 200 different diagnoses; these medical problems coupled with social, educational, financial, and other family concerns are difficult to address, and there is no single service delivery system to meet the needs of every person. We work with the individuals and their families as our clients. We coordinate and integrate an array of community services. Some clients need the services of both mental health and mental retardation. Professionals are not provided the practical experience to enable them to cross service delivery systems.
>   —Margaret Mikol, Sick Kids Need Involved People of New York, personal
>   communication, May 4, 1994

*O*ur agency has always served adolescent females. They come from a variety of situations. They might be "system" kids who are in the custody of the state and need temporary shelter, they might be having difficulties at home so they would be private placements in which their parent or guardian places them for crisis intervention, they could be placements through the juvenile court, or they could come right off the street having been on the run or homeless. We are a short-term facility in which the length of stay varies depending on the type of placement— private placements can stay up to 14 days and "system" kids placed by the state can stay up to 30 days. We provide as much as possible in the way of services, including individual and family therapy, psychoeducational groups, a one-time mental health assessment completed by the staff psychologist, along with basic needs of food, shelter and clothing.
>   —Stacie Newberry, Marion Hall Emergency Shelter, St. Louis, MO, personal
>   communication, November 15, 2001

*I*ntensive Case Management Program is one that is set up to service long-term clients, people who have needs that you just can't meet in two or three months. . . . We do about everything there is to do to try to aid and help with our clients. . . . Normalization is real important in mainstreaming them into the community . . . and trying to erase the stigma of mental illness . . . providing the necessary services that we feel that will assist them, from daily living skills to transportation, health needs, and medications.
>   —Jana Berry Morgan and Paula Hudson, Helen Ross McNabb Center, Inc.,
>   Knoxville, TN, personal communication, April 27, 1995

The preceding quotations are the words of case managers who are involved in the delivery of human services. This chapter introduces you to the subject and presents a model of case management that guides many helping professionals who work in human service delivery. Focus your reading and study on the following objectives.

CASE MANAGEMENT DEFINED
- Describe the context in which human service delivery occurs today.
- Differentiate between traditional case management and case management today.

THE PROCESS OF CASE MANAGEMENT
- List the three phases of case management.
- Identify the two activities of the assessment phase.
- Illustrate the role of data gathering in assessment and planning.
- Describe the helper's role in service coordination.

THREE COMPONENTS OF CASE MANAGEMENT
- Define case review and list its benefits.
- Support the need for documentation and report writing in case management.
- Trace the client's participation in the three phases of case management.

PRINCIPLES AND GOALS OF CASE MANAGEMENT
- List the principles and goals that guide the case management process.
- Describe how each principle influences the delivery of services.

Case Management Defined

The world in which case managers function is changing rapidly. Because of client tracking systems, the electronic transfer of records, dual-diagnosis clients, limited resources, and rapid communication capabilities, current service delivery is vastly different from that of just a few years ago. One result is that the time between policy development and implementation is much shorter. Another is that many human service agencies and organizations have chosen to limit the services they provide. More and more, case managers need skills in teamwork, networking, referral, and coordination in order to obtain the services clients need. All this takes place in a constellation of service providers that continues to grow and change.

In addition, service delivery is affected by a political climate in which the role of government in human services comes under close scrutiny. How involved should government be in meeting human needs? What is its role? What is the proper relationship between state and federal governments? As these questions are examined and debated, case managers sometimes find themselves working under a cloud of uncertainty that influences the work they do, their professional identity, and their professional development.

The quotations that introduced this chapter share a common theme: All three situations require providing and coordinating services for the individuals and families served. Margaret Mikol directs an agency that provides intensive case management to children and families with complex medical problems. In this agency, the case management process begins as early as the diagnosis of a medical problem and can be terminated once clients are back home and able to manage their own care. Clients are supported by an assessment, planning, and coordination process. There is a continuous evaluation of both client needs and the effectiveness of the care provided. Since the ultimate goal is for the family to manage its own case, all plans and services focus on and build on family strengths.

The services provided by the Marion Hall Emergency Shelter are different. Its primary responsibility is to provide housing, assessment, and counseling for two weeks; the staff then make appropriate referrals. Although contact is short-term, the girls receive intensive physical and psychological care, participate in determining their own treatment plan, and receive shelter and nutritious food. The treatment plan is based on their needs, strengths, and interests. Accountability means developing plans based on the girls' priorities as established on the day they arrive.

Paula Hudson and Jana Berry Morgan work in an agency that provides long-term managed care for people with mental illness. Rarely do they close a case. People with severe mental illness who reside in the community require service coordination that is long term, closely monitored, and supportive. The agency's commitment to these clients is to assess their needs periodically and adjust plans and provide services accordingly. Often this agency is the only lifeline for these adults. Since the agency maintains a long-term relationship with clients, its staff develop ways to update assessments and service plans.

These diverse examples illustrate service delivery today. As you can see, the care varies from agency to agency, from helper to helper, and from client to client. One element each example has in common is the use of case management to coordinate and deliver services, moving an individual through the service delivery process from intake to closure.

Traditional Case Management

To define case management, it is helpful to look at the ways in which case management was traditionally regarded. In mental health service delivery in the 1970s, case management was a necessary component of service delivery because clients with complex needs required multiple services. Case management was a process linking clients to services that began with assessment and continued through intervention. In the 1980s, there was a shift in the focus of case management. Many professionals and clients objected to the use of the word *manage* because its definition is "to control and direct; to handle; to treat" (*Webster's New Collegiate Dictionary*, 1973, p. 697). This language did not seem to reflect a commitment to client involvement or empowerment. Terms such as *service coordination* and *care coordination* were considered to indicate more completely these new goals of case management. Many believed that the term *service coordination* more accurately represented the primary work of the case management process—linking the client to services and monitoring progress. Jackson, Finkler, and Robinson (1992) describe the development of the term *care coordination* during their work with Project Continuity, which facilitated care for infants and toddlers who required repeated hospitalization and who qualified for intervention under Public Law 94-457, the Individuals with Disabilities Act of 1986.

> Over the course of this project, the term *care coordination* evolved from what is popularly described as case management. Staff expressed dissatisfaction with the case management term because they did not feel families should be viewed as cases needing to be managed. Therefore, the project changed the description to care coordinator, which reflects the role as coordinator of care services for the child and family. (p. 224)

Case Management Today

In the late 1990s and early 2000s, many effective case managers assume the dual role of linking and monitoring services and providing direct services. In many instances, this dual role is called intensive case management and reflects the time and financial resources committed to the client.

We have conducted numerous interviews with service providers who are doing case management, and some indicate a preference for terms other than case management and case manager in describing their jobs and job titles (McClam & Woodside, 1994). Three primary objections to these terms surfaced. One is that the practitioners find it objectionable to think of clients as cases. A second relates to the resentment clients may feel at being managed. Third, these helpers believe that they do more than case management. Many of the helpers interviewed did refer to themselves as case managers, but not necessarily in the traditional sense of the term.

What has emerged today is a broader perspective on service delivery, one that encompasses traditional case management, as well as case management with a broader focus. In some situations, it includes case management with a new focus. **Case management** is a creative and collaborative process, involving skills in assessment, consulting, teaching, modeling, and advocacy that aim to enhance the optimum social functioning of the client served (Mullahy, 1998; Sullivan, Wolk, & Hartmann, 1992). Note that it includes the dual role of coordinating and providing direct service. **Case managers** are the helping professionals who perform the responsibilities of case management. The goal of case managers is to help those who need assistance to manage their own lives and to support them when expertise is needed or a crisis occurs. These professionals gather information, make assessments, and monitor services. They find themselves working with other professionals, arranging for services from other agencies, serving as advocates for their clients, and monitoring resource allocation and quality assurance. They also provide direct services, describing their responsibilities broadly as doing whatever is necessary to help the client.

The evidence is clear that case management is more a part of service delivery than ever before. In fact, case management is defined and mandated through federal legislation, has become part of the services offered by insurance companies, and is now accepted by helping professionals as a way to serve long-term clients who have multiple problems.

The diversity of professionals with case management responsibilities is reflected in the many job titles they carry: case manager, intensive case manager, service coordinator, counselor, social worker, service provider, care coordinator, caseworker, and liaison worker. In some cases, these professionals provide services themselves; in others, they coordinate services or manage them. Increasingly, they are assuming new responsibilities, such as cost containment and budget management. There is little agreement about what to call those they serve, but most frequently they talk about *clients*, *individuals*, or *participants*.

The diversity of job titles, the range of individuals and groups served, and the variety of job responsibilities are all indications that service delivery is changing. This text explores case management as a complex, evolving, and diverse process. You will review traditional case management, learn about the new ways in which case management is being applied, and explore the new roles and responsibilities given to helpers.

One of the important ways of learning about case management is through the voices of helping professionals themselves, as in the many concrete examples in this book. As you read, note their different job titles, roles, responsibilities, service delivery methods, and terminology. The examples that illustrate concepts and principles generally use the terminology of the particular setting involved. When a case or example does not define the terminology, the term *case management* will be used to mean the responsibilities of both service provision (e.g., counseling) and service coordination (e.g., arranging for services from others). The term also refers to the management skills needed to move a case from intake to closure. In referring to the service provider, the term *case manager* will mean the professional who performs the tasks of case management.

The section that follows introduces the process of case management and its three phases. The case of Roy Roger Johnson illustrates each phase.

 ## The Process of Case Management

The three phases of case management are assessment, planning, and implementation. (See Figure 1.1.) Human service delivery has become increasingly complex in terms of the number of organizations involved, government regulations, policy guidelines, accountability, and clients with multiple problems. Therefore, the case manager needs an extensive repertoire of knowledge, skills, techniques, and strategies.

Let's see how these phases occur in three different settings. Thom Prassa is a case manager at the East Tennessee Community Health Agency, which has initial responsibility for all children who come through the juvenile court system. He spends much of his time making assessments of young people who are transferred to correctional facilities in the state and the community. For him, assessment is complex and multifaceted. He describes it this way.

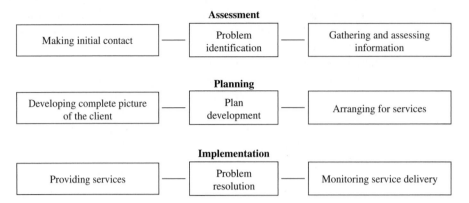

Figure 1.1   The process of case management

I will gather every bit of information there is about a child: school records, medical records, prior psychological evaluations. I might arrange for a psychological evaluation or interview a guidance counselor and parents. Do home visits. If the child is referred to us after coming into custody, then one of my main responsibilities is to get that child a physical. (Personal communication, July 10, 1994)

Yolanda Vega, Director of Agency Services at Casita Maria Settlement House in the Bronx, describes the process of planning how her staff will provide services to clients.

It is very important to take one step at a time. Narcotics Anonymous and Alcoholics Anonymous talk about steps. Our clients want to do so much all at once. You have to talk with them and convince them that the right way to go is one day at a time. You do not try to battle all these demons at once. Sometimes the worker might have high expectations and the client is really in a rush to get there. But oftentimes that does not work. (Personal communication, May 4, 1994)

Another social worker serves as a case manager in the emergency room of a large metropolitan hospital. He provides frontline assessment and referral for treatment for emergency admissions from various sources. Finding a place for patients to stay represents the implementation phase.

The social worker is the person who works with the patient initially. Patients are generally brought in by the police, usually involuntarily. You do all the paperwork, all legal paperwork. You get all the paperwork back that commits them here . . . the paperwork is very important especially if the patient is hostile. Some are pure emergencies that go away in 24 hours . . . still it is the social worker's job to find them somewhere to go. (Walter Rae, Houston, TX , personal communication, December 16, 1999).

As you can see, the responsibilities at each phase vary, depending on the setting and the case manager's job description. It is important to understand that the three phases represent the *flow* of case management rather than rigidly defined steps to successful case closure. An activity that occurs in the first phase (e.g., the information gathering that Thom Prassa does) may also appear in the second or third phases as in Yolanda Vega's planning and the hospital social worker's referral. Other key components in effective case management appear throughout the process, including case review, report writing and documentation, and client participation. Ultimately, the goal of case management, stated earlier, is to empower clients to manage their own lives as well as they are able. The case of Roy Roger Johnson shows how this happens.

Roy Roger Johnson is a real person, but his name and other identifying information have been changed. The case as presented here is an accurate account of Roy's experience with the human service delivery system and the case management process. His case exemplifies the three phases of case management. The agency that served Roy uses the terms *counselor* and *client*. The following background information will help you follow his case through assessment, planning, and implementation.

Roy referred himself for services after suffering a back injury at work. He was 29 years old and had been employed for five years as a plumber's assistant; he hurt his back lifting plumbing materials. After back surgery, he wanted help finding work. Although he had received a settlement, he knew that the money would not last long, especially since he had contracted to have a house built. He heard about the agency from a friend who knew someone who had received services there and was now working. The agency helps people with disabilities that limit the kind of work they can do. An important consideration in accepting a person for services at the agency is determining whether services will enable that person to return to work. Roy's case was opened at the agency; we will follow it to closure.

Assessment

The **assessment** phase of case management is the diagnostic study of the client and the client's environment. It involves initial contact with an applicant as well as gathering and assessing information. These two activities focus on evaluating the need or request for services, assessing their appropriateness, and determining eligibility for services. Until eligibility is established, the individual is considered an applicant. When eligibility criteria have been met, the appropriateness of service is determined, and the individual is accepted for service, then he or she becomes a client.

THE INITIAL CONTACT

The initial contact is the starting point for gathering and assessing information about the applicant to establish eligibility and evaluate the need for services. In most organizations, the data gathered during the initial contact is basic and demographic: age, marital status, educational level, employment information, and the like. Other information may be obtained to provide detail about aspects of the

---

APPLICATION FOR SERVICES

Part 1

A. Office No. ☐☐   B. Counselor No. ☐☐☐   C. Client SSN ⑴⑵⑶-⑷⑸-⑹⑺⑻⑼

D. Review Date ☐☐-☐☐-☐☐   E. Referral Date ☐☐-☐☐-☐☐

F. Name____*Johnson,*____   ____*Roy*____   ____*Roger*____
                Last   First   Middle

G. Address __*Rt. #51*__ __*Centerville*__   ☐☐ *TN* ⑴⑴⑵⑵⑵
                Street   City   County   State   Zip

H. Phone No.⑺⑴⑴-⑸⑶⑴-⑺⑴⑴⑷ Directions to home:_____

I. Birthdate ⑴⑺-⑴⑹-⑹⑴ Age *29* Place of birth: *Michigan   East Lansing*

J. Referral Source ☐☐ ___*self*___

K. Disability ☐☐☐ ___*bad back*___   L. Sex Ⓜ M-Male F-Female

L. Cause of Disability: *accident*_____ Age at beginning of disability: *28*

M. How does disability limit activities? *walking, standing, pushing, pulling, lifting*

N. Other physical or mental problems: *none*_____

O. Have you previously received Services?   Yes____ No *✓* State_____ Date_____

Part 2

A. Race ☐ 1-White 2-Black 3-American Indian or Alaskan Native 4-Asian or Pacific Islander

B. Highest grade of school completed ⑴⑵ (MR=99) Year____ Name & Address:_____

Other Training:_____

C. Martial status ⑸ 1-M,2-W,3-Div,4-Sep,5-NM D. No. of Dep☐☐ E. Hisp. Origin: Yes☐No☑ F. Vet: Yes☐ No☑

G. Wk status ☐   H. Wkly Earn ☐☐☐   I. Hrs Worked ☐☐   J. Primary source of support ☐ K. Institution ☐☐

L. Public Assistance/Support:

SSI-Aged	Yes ☐ No ☐ Amount_____	SSDI	Yes ☐ No ☑ Amount_____	
SSI-Blind	Yes ☐ No ☐ Amount_____	VA Disability	Yes ☐ No ☑ Amount_____	
SSI-Disabled	Yes ☐ No ☐ Amount_____	Other Disability	Yes ☐ No ☑ Amount_____	
AFDC	Yes ☐ No ☐ Amount_____	Other PA/PS	Yes ☐ No ☑ Amount_____	
Gen. Ass't	Yes ☐ No ☐ Amount_____	Total monthly amount (Nearest $)☐☐☐ (none 000)		

M. Medical Insurance Coverage: Yes ☐ No ☐ Name & Type Coverage:_____

_____ Coverage ID No._____

N. Availability of Medical Insurance through Client's Employment for a Salary or Wages at Application: ☐
   0 - Insurance not available   1 - Insurance is available   2 - Client not working at application

Figure 1.2   Application for services

client's life, such as medical evaluations, social histories, educational reports, and references from employers.

Roy was self-referred to the agency. He initiated contact by telephoning for an appointment. Fortunately, a counselor was able to see him that week, so he made an appointment for May 24 at 10:30. He was sent a brochure about the agency and a confirmation of his appointment. When he arrived at the agency,

Name & Address of Family Physician: *Dr. Alderman*

Name & Address of Physicians and Dates Seen for Disability: *Dr. Alderman   06/16/XX*

Date Last Hospitalized: *May XX*   Hospital: *Centerville*   Reason: *Surgery*

Do you wear any type artificial appliance? Yes_____ No _✓_ Type:_____

Family Members Living in Home:

NAME	AGE	RELATION	EDUCATION	JOB	MONTHLY WAGE
Separate family entity					

Past Work Record: [LIST LAST JOB FIRST]

EMPLOYER	ADDRESS	JOB TITLE	DATES	WEEKLY WAGE $	WHY DID YOU LEAVE THIS JOB?
Memorial Hosp.	*Centerville*	*Plumber*	*2 mos. XX*	$ *7.50/hr*	*hurt on job*
Rock City	*Mechanical*			$	
				$	
		Construction work all		$ *s/c*	
				$	

Type or Area of Vocational Interest: *undecided*

List 2 persons (Other than listed in home) who would always know your address:

Name: *Terry Jones*   Address:_____   Phone: *987-6543*

Name: *Mae Johnson*   Address:_____   Phone: *123-4567*

APPLICATION FOR REHABILITATION SERVICES*

I hereby make application to receive services I may be eligible to receive. I understand that the information contained in my record will not be disclosed, other than in the administration of the program unless written consent is obtained from me or my parent/guardian. Also, I hereby request and authorize any person(s), agency or institution to release to the agency any medical, social, psychological, vocational and/or financial information they may have or may receive, pertaining to me.

Signature of Client *Roy R Johnson*_____   Date ☐☐-☐☐-☐☐

Signature of Parent/Guardian (if required)_____   Date_____

Signature of Counselor *Tom Chapman*_____   Date *5/24/XX*

*To be explained at the time of initial interview   Review Date ☐☐-☐☐-☐☐

Figure 1.2   (*Continued*)

Roy completed an application for services. (See Figure 1.2.) The agency believes the applicant should supply the information in this initial information gathering. He was able to complete it without too much trouble, although he wasn't sure how to answer the question about where he had heard about the agency. He didn't know the name of his friend's friend. The receptionist helpfully told him to write in "self-referral." She suggested that he leave any questions blank if he wasn't sure about the response. She also asked him not to

---

ORTHOPEDIC ASSOCIATES

200 W. MAIN STREET
DOUGLAS, USA 12345-6789

June 16, XXXX

Mr. Jefferson Maupin
Attorney at Law
215 Fourth Street
Douglas, USA 12345-6788

Re: Roy Roger Johnson

Dear Mr. Maupin:

HISTORY: Mr. Roy Johnson presented to my office on April 12, with a history of an injury to the lumbar spine, which occurred at work on March 15. He reported that he was lifting an object weighing approximately 60 lb. at the time of onset. Prior to that event, he had not experienced significant low back or lower extremity pain for the preceding four years. The patient does have a significant history, including previous lumbar spine surgery. The surgical procedure was performed in July XXXX, and included a bilateral L4 laminectomy and diskectomy with a left L5 foraminotomy. When the patient presented to my office on this occasion, he complained of right lower extremity pain rather than left lower extremity pain. My initial disposition was to refer the patient to Physical Therapy and ask him not to work for two weeks.

The patient returned to my office on April 25, at which time he continued to complain of severe low back and right lower extremity pain. On that occasion, an MRI study was ordered, which revealed evidence of a herniated disc at the L4 level primarily on the right consistent with the patient's right lower extremity pain.

Mr. Johnson was admitted to the hospital for definitive surgery and, on May 4, he underwent a right L4 diskectomy with the operating micro-

---

Figure 1.3   Dr. Alderman's letter

sign the application until he had met with a counselor. She stated that each counselor liked to explain the paragraph at the end of the application in order to make sure that applicants understood the implications of applying for services and the conditions that apply to the release of any client information. She would later transfer this information to the agency's client database.

scope. The procedure was more difficult because of previous surgical scar tissue. At surgery, the patient had several free extruded disc fragments consistent with acute lumbar disc herniation.

Since surgery, the patient has returned to the office on several occasions for routine follow-up. He has also been attending Physical Therapy for a routine postoperative physical therapy program.

DISCUSSION: This patient sustained an acute lumbar disc herniation on March 15, as evidenced by acute onset of back and leg pain documented by MRI study and by positive surgical findings. Because the patient has a history of significant previous lumbar disc disease, it is my opinion that he will undergo a gradual recovery and that he will be left with a disability to the body as a whole as a result of the injury described above of 10 percent to the body as a whole. This disability rating does not reflect his preexisting disability. The disability rating is based on anatomical findings and is in accordance with the AMA guidelines.

Sincerely,

MF Alderman

Marvin F. Alderman, M.D.

MFA:bj

Figure 1.3   *(Continued)*

Roy had brought a copy of a letter prepared by his orthopedic surgeon, Dr. Alderman, for his attorney a year earlier. (See Figure 1.3.) Dr. Alderman had expressed the opinion that Roy would be left with a disability of 10% as a result of the injury. Dr. Alderman was also careful to clarify that Roy's condition did not reflect a preexisting disability even though he had suffered back problems previously. Tom Chapman, the counselor who saw Roy, made a copy of the letter and returned Roy's copy to him.

During the initial contact, the case manager determines who the applicant is, begins to establish a relationship, and takes care of such routine matters as filling out the initial intake form. An important part of getting to know the applicant is learning about the individual's previous experiences with helping, his or her strengths, his or her perception of the presenting problem, the referral source, and the applicant's expectations. As these matters are discussed, the case manager uses appropriate verbal and nonverbal communication skills to establish rapport with the applicant. (These skills will be discussed in Chapter 5.) Skillful use of interviewing techniques facilitates the gathering of information and puts the applicant at ease. The counselor makes the point at the conference that the client is considered an expert and that self-reported information is very important. By providing information about routine matters, the case manager demystifies the process for the applicant and makes him or her more comfortable in the agency setting. Some of the routine matters addressed during the initial meeting are completing forms, gathering insurance information, outlining the purpose and services of the agency, giving assurances of confidentiality, and obtaining information releases.

Documentation records the initial contact. In the agency Roy went to, case managers fill out a Counselor's Page (Figure 1.4), which describes the initial meeting, and a Client Master Record (Figure 1.5). The Client Master Record provides basic information about the client, his or her sources of support, and employment. Its format was designed so that data could easily be entered into the computer, thus simplifying the agency's recordkeeping. At this point, Roy was still considered an applicant for services in accordance with agency guidelines.

Although Dr. Alderman's letter provided helpful information about Roy's presenting problem, agency guidelines stated that all applicants must have a physical examination by a physician on the agency's approved list. Mr. Chapman also felt that a psychological evaluation would provide important information about Roy's mental capabilities. He discussed both of these with Roy, who was eager to get started. As Roy prepared to leave, Mr. Chapman explained that it would take time to process the forms and review his application for services. He would be in touch with Roy very soon, explaining the next steps. (See Figure 1.6.)

GATHERING AND ASSESSING INFORMATION

If the applicant is accepted for services, the client and the case manager will become partners in reaching the goals that are established. Therefore, as they work through the initial information gathering and routine agency matters, it is important that they identify and clarify their respective roles, as well as their expectations for each other and the agency. From the first contact, client participation

NAME: <u>Roy R. Johnson</u>   Date _____

Mr. Johnson is a 29-year-old referral that has an orthopedic back problem and brought with him a doctor's report from the first accident as well as the second. The client has limitations in most of his daily living activities. The client has just settled with unemployment for $40,000 and is presently building a home, which will deplete this very soon. The client has a 12th-grade education and stated he had been through 2 college quarters. The client was pleasant and well mannered and answered the questions without any problems.

The rules, regulations, and time limits were explained and understood by the client as well as the order of selection. The client is seeking possible training, but really is undecided about what he can and cannot do. He would meet the economic guidelines for certain services at this time and, if the statement is correct, for all services, once his settlement is depleted.

The client was given a functional limitation sheet for Dr. Alderman, which he is to return to this office. We will sponsor a general medical and psychological evaluation with Barbara Hillman. We may put him in a vocational evaluation at the TVTC. We will place this case in status 02 as of this date.

TC:bj

Figure 1.4   Counselor's Page

and service coordination are critical components in the success of the process. The case manager must make clear that the client is to be involved in all phases of the process. A skillful case manager makes sure that client involvement begins during the initial meeting.

In Roy's case, the counselor reviewed the application with him. There were some blanks on the application, and they completed them together. Roy had not

CLIENT MASTER RECORD

APPLICANT STATUS 02

A. Office No. ☐☐

B. Counselor No. ②⑤⑨

C. Client No. ⓪②③■④⑤■⑥⑦③⑨

D. Client Name *Johnson, Roy R.*
LAST FIRST MIDDLE

E. Review Date: ☐☐■☐☐■☐☐

F. Date of: Status 02 ⓪②■⓪②■⑨⑨

P. Work Status
1. Wage or salaried worker-competitive labor market
2. Wage or salaried worker-sheltered workshop
3. Self-employed-except state agency managed business
4. State agency-managed business enterprise
5. Homemaker
6. Unpaid family worker
7. Not working-student
⑧. Not working-other
9. Trainee or worker (non-competitive labor marker)

G. Race
①. White
2. Black
3. American Indian or Alaskan Native
4. Asian or Pacific Islander

Q. Hours Worked ⓪⓪

H. Highest Grade Of School Completed ①②
(M.R. = 99)

R. Institution
⓪⓪ Not in institution
01 - Public Mental Hospital
02 - Private Mental Hospital
03 - Psychiatric inpatient unit of General Hospital
04 - Community Mental Health Center-inpatient
05 - Public institution for the mentally retarded
06 - Private institution for the mentally retarded
07 - Alcoholism treatment center

I. Marital Status
1. Married
2. Widowed
3. Divorced
4. Separated
⑤. Never married

08 - Drug abuse treatment center
09 - School and other institution for the blind
10 - School and other institution for the deaf

J. Number of Dependents ⓪②

K. Hispanic Origin Yes ☐ No ☒

11 - General Hospital
12 - Hospital or specialized facility for chronic illness
13 - Institution for aged
14 - Halfway House

L. Veteran Yes ☐ No ☒

15 - Correctional Institution - adult
16 - Correctional Institution - juvenile
17 - Other institutions and special living arrangements including group home

M. Occupation (Title) *none*

Occupation Code ③②①⓪①③⑨

S. Public Assistance/Public Support
⓪. Not on public assistance
1 - SSI - aged
2 - SSI - blind

N. Weekly Earnings (Nearest $) $☐☐☐None☒000

3 - SSI-disabled
4 - AFDC
5 - GA only
6 - SSDI

O. Primary Source of Support
A. Current earnings, interest, dividends, rent
B. Family and friends
C. Private relief agency
D. Public assistance, at least partly with Federal funds

7 - Veterans' disability benefits
8 - Other disability benefits
9 - All other PA/PS payments

E. Public assistance, without Federal funds (General Assistance only)
F. Public institution-tax supported
G. Workmen's compensation
H. Social Security Disability Insurance benefits
J. All other public sources
Ⓚ. Annuity or other non-disability insurance benefits (private insurance)
L. All other sources of support

T. Public Assistance Monthly Amount

(Nearest $) $☐☐☐ None☒000

U. Weeks Unemployed ②⑤

Tom Chapman
COUNSELOR'S SIGNATURE

6/25/xx
DATE

Figure 1.5 Client Master Record

been sure how to respond to the questions about primary source of support and members of his household. As Roy elaborated on his family situation, the counselor completed these items. Roy felt positive about his interactions with Tom Chapman because Tom listened to what he said, accepted his explanations, and showed insight, empathy, and good humor.

July 20, XXXX
Roy R. Johnson
Rt. 51
Centerville, USA 12345

Dear Mr. Johnson:

We have scheduled the following appointment(s) for you: We have authorized a general physical for you with Dr. Jones, Suite 201, Physicians Office Building, 172 Lake Road. Please call 589-2111 to schedule the appointment. We have also authorized a psychological evaluation with Barbara Hillman. She will call you to schedule an appointment.

Please make every effort to keep the appointment(s) that we have scheduled. If, however, you will be unable to keep the appointment(s) please contact this office prior to the date of the appointment. Our phone number is 596-5120. Your cooperation is appreciated. If you have any questions, please feel free to contact us.

Sincerely,

Tom Chapman

Rehabilitation Counselor

TC:bj

Figure 1.6   Tom Chapman's memo

In gathering data, the case manager must determine what types of information are needed to establish eligibility and to evaluate the need for services. Once the types of information are identified, the case manager decides on appropriate sources of information and data collection methods. His or her next task is making sense of the information that has been gathered. In these tasks, assessment is involved: The case manager addresses the relevance and validity of data and pieces together information about problem identification, eligibility for services, appropriateness of services, plan development, service provision, and outcomes evaluation. During this process the case manager checks and rechecks the accuracy of the data, continually asking, "Does the data provide a consistent picture of the client?"

Client participation continues to play an important role throughout the information-gathering and assessment activities. In many cases, the client is the primary source of information, giving historical data, perceptions about the presenting problem, and desired outcomes. The client also participates as an evaluator of information, agreeing with or challenging information from other sources. This participation establishes the atmosphere to foster future client empowerment.

The counselor needed other information before a certification of eligibility could be written. In addition to Dr. Alderman's letter, a general medical examination, and a psychological evaluation, the counselor requested a period of vocational evaluation at a regional center that assesses people's vocational capabilities, interests, and aptitudes. Tom Chapman had worked with all these professionals before, so he followed up the written reports he received with further conversations and consultations. Following a two-week period at the vocational center, the evaluators met with Roy and Mr. Chapman to discuss his performance and make recommendations for vocational objectives. When the report was completed, Mr. Chapman and Roy met several times to review information, identify possibilities, and discuss the choices available to Roy. Mr. Chapman's knowledge of career counseling served him well as he and Roy discussed the future. Unfortunately, an unforeseen complication occurred, delaying the delivery of services. Tom Chapman changed districts, and another counselor, Susan Fields, assumed his caseload. Meanwhile, Roy moved to another town to attend college. Although he was still in the same state, Roy was now about 200 miles from his counselor. While Roy was attending his first semester at school in January, Ms. Fields completed a certificate of eligibility for him. (See Figure 1.7.) This meant that he was accepted as a client of the agency and could now receive services. In May, his case was transferred to another counselor (his third) in the town where he lived and attended college.

Planning

The second phase of case management is **planning**, which is the process of determining future service delivery in an organized way. When planning begins, the agency has usually accepted the applicant for services. The individual has met the eligibility criteria and is now a client of the agency. During this planning process, the counselor and the client turn their attention to developing a service plan and arranging for service delivery. Client participation continues to be important as desired outcomes are identified, services suggested, and the need for additional information determined. The actual plan addresses what services will

Contact Report and Case Memorandum

Name Roy R. Johnson   Date of contact: 1/5/xx

District 31   Counselor: S. B. Fields

Status: Last report 02   This report 10   Date of change 1/5/xx

Certification of Eligibility

Disability: The client's primary disability is a back condition. The enclosed general medical exam and orthopedic report from Dr. Porter substantiate this. At present, there are no secondary disabling conditions identified.

Functional limitations/vocational handicap: The client has had long-term treatment for his back condition. He is able to lift only 20 lbs; is restricted to walking, standing, sitting, and occasionally bending; and has restrictions on his lower extremities. The client's vocational handicap restricts him from numerous job areas, such as construction work and truck driving.

Reasonable expectations: I feel that the client can benefit from an educational program before going to selective job placement.

Severely disabled: The client presently meets our guidelines for severe disability, as he will require multiple services over an extended period of time. He clearly has long-term impairment.

Priority category: The client is being assigned to Priority Category I.

Susan Fields   1/5/xx
Counselor's signature   Date

bj

Figure 1.7   Certificate of eligibility

be provided and how they will be arranged, what outcomes are expected, and how success will be evaluated.

A plan for services may call for the collection of additional information to round out the agency's knowledge of the client. Some case managers suggest that the service delivery process is like a jigsaw puzzle, with each piece of information providing another clue to the big picture. During this stage, the case manager may realize that a social history, a psychological evaluation, a medical evaluation, or educational information might provide the missing pieces. The plan identifies what services are needed, who will provide them, and when they will be given. The case manager must then make the appropriate arrangements for the services.

During the assessment phase, Tom Chapman did a comprehensive job of gathering information about Roy. When Roy was accepted for services, the task facing him and his new counselor was to develop a plan of services. (See Figure 1.8.) Clarity and succinctness characterize the service plan, which the counselor and the client complete together, emphasizing the client's input in the process. The plan lists each objective, the services needed to reach that objective, and the method or methods of checking progress.

Suppose that Tom Chapman had believed that a psychological evaluation was unnecessary and had been able to establish eligibility solely on the basis of the medical and vocational evaluations. Susan Fields, the new counselor, might find that a psychological evaluation would be beneficial, especially since the agency was contemplating providing tuition and support for a college education. One objective of the plan would then be to provide a psychological evaluation of the client. This is an example of continuing to gather data during the planning phase, as well as continuing to assess the reliability and validity of the data.

Roy's plan indicates that he is eligible for services and meets agency criteria. His program objective, business communications, was established as a result of evaluation services, counseling sessions with Mr. Chapman, and Roy's stated vocational interests. The three stated intermediate objectives will help Roy achieve the program objective. The plan also provides a place to identify the responsibilities of Roy and of the agency in carrying out the plan. Note that this agency takes very seriously the participation of the client in the development of the plan, even asking that the client sign it, as well as the counselor. There is one additional part of the plan to be signed by the client. Because one of Roy's objectives is to attend college, which involves a significant expenditure on the agency's part, Roy has also signed a Student Letter of Understanding that further describes his responsibilities. In a sense, this letter is a contract between the student and the agency; the counselor signs it as the agency's representative. The letter also states that this agreement is valid as determined by federal and state regulations.

Once the plan is completed, the counselor begins to arrange for the provision of services. He or she must review the established network of service providers. Experienced case managers know who provides what services and who does the best work. Nonetheless, they should continue to develop their networks. For beginning helpers, the challenge is to develop their own networks: identifying their own resources and building their own files of contacts, agencies and services. Chapter 6 provides information about developing, maintaining, and evaluating a network of community resources.

Implementation

The third phase of case management is **implementation**, when the service plan is carried out and evaluated. It starts when service delivery begins, and the case

SERVICE PLAN

1. NAME ___JOHNSON, ROY_____ PROGRAM TYPE ☒ INITIAL ☐ AMENDMENT

2. YOU ARE ELIGIBLE FOR: ☒ VOCATIONAL REHABILITATION SERVICES ☐ EXTENDED EVALUATION SERVICES ☐ PAST EMPLOYMENT SERVICES

    BECAUSE: ☒ A. YOU HAVE A PHYSICAL OR MENTAL DISABILITY WHICH CONSTITUTES A SUBSTANTIAL HANDICAP TO EMPLOYMENT AND:

        ☒ B. YOU CAN REASONABLY BE EXPECTED TO BENEFIT IN TERMS OF EMPLOYABILITY FROM SERVICES.

        ☐ C. IT CANNOT BE DETERMINED WHETHER OR NOT YOU CAN BENEFIT IN TERMS OF EMPLOYABILITY FROM
            REHABILITATION SERVICES.

        ☐ D. POST EMPLOYMENT SERVICES ARE NEEDED FOR YOU TO MAINTAIN EMPLOYMENT.

3. PROGRAM OBJECTIVE:_Business Communications___ANTICIPATED DATE OF ACHIEVEMENT: MONTH_1_YEAR_XX_
                                      ESTIMATED DATES TO REACH OBJECTIVE & RECEIVE SERVICES
4. INTERMEDIATE OBJECTIVE, SERVICES METHODS OF CHECKING PROGRESS.

	RESPONSIBILITY	FROM	TO
OBJECTIVE_To correct physical impairment so that client might_			
SERVICES _reach vocational objective_	Client	1/XX	1/XX
Possible office visit with the doctor			

METHOD OF CHECKING PROGRESS___Medical information_

	RESPONSIBILITY	FROM	TO
OBJECTIVE _To provide background information and educational skills_			
SERVICES _so that client might reach vocational objective_	VR	1/XX	1/XX
A. Tuition/UT Knoxville		1/XX	1/XX
B. Miscellaneous Educational Expenditures		1/XX	1/XX

METHOD OF CHECKING PROGRESS ___R-11, Grade Reports_

	RESPONSIBILITY	FROM	TO
OBJECTIVE_To follow client's progress and develop plan amendment if needed so that client might_			
SERVICES _reach objective_	Client	1/XX	1/XX
A. Possible RP-B		1/XX	1/XX
B. Client/Counselor Contacts		1/XX	1/XX

METHOD OF CHECKING PROGRESS_R-11_

5. CLIENT OR FAMILY AND AGENCY RESPONSIBILITIES AND CONDITIONS: I. Client is responsible to maintain contact with counselor twice each semester by mail, phone or in person. II. Client is responsible to furnish VR Counselor with a copy of grades at the end of each term. III. Client is responsible to maintain an average load of classes and average grades throughout his program. IV. Client is responsible to file for any similar benefits which might help him pay for his program. V. Client is responsible to furnish VR counselor with a resume and a list of potential employers to interview with during the first part of his senior year. VI. Client is responsible to notify counselor of any significant change of address, health, phones number of financial status.

6. CLIENT'S VIEW OF PROGRAM ___The client and I have discussed the services necessary to help him reach his vocational objective and we are in mutual agreement with his plan._

I HAVE PARTICIPATED IN THE DEVELOPMENT OF THIS PROGRAM AND I UNDERSTAND IT.
I UNDERSTAND AND ACCEPT THE STATEMENT OF UNDERSTANDING WHICH HAS BEEN EXPLAINED TO ME.

Roy Johnson   5-6-XX   _Susan Fields_   5/6/XX
Client's Signature   Date   Supervisor Signature   Date

Figure 1.8   Service plan

manager's task becomes either providing services or overseeing service delivery and assessing the quality of the service delivery. He or she addresses the questions of who provides each service, how to monitor implementation, how to work with other professionals, and how to evaluate outcomes.

In general, the approval of a supervisor is needed before services can be delivered, particularly when funds will be expended. Many agencies, in fact, have a cap (a fee limit) for particular services. In addition, a written rationale is often required to justify the service and the funds. As resources become increasingly limited, agencies redouble their efforts to contain the costs of service delivery. In Roy's case, the agency's commitment to pay his college tuition represented a significant expenditure. Susan Fields submitted the plan and a written rationale to the agency's statewide central office for approval.

Who provides services to clients? The answer to this question often depends on the nature of the agency. Some are full-service operations that offer a client whatever services are needed in-house, as described in Chapter 3. As a rule, however, the client does not receive all services from a single worker or agency. It is usually necessary for him or her to go to other agencies or organizations for needed services. This makes it essential for the case manager to possess referral skills, knowledge of the client's capabilities, and information about community resources.

No doubt you remember that Roy's first counselor, Tom Chapman, arranged for a psychological evaluation. Many agencies like Tom's have so many clients needing psychological evaluations that they hire a staff psychologist to do in-house evaluations of applicants and clients. Other agencies simply contract with individuals—in this case, licensed psychological examiners or licensed psychologists—or with other agencies to provide the service. Whatever the situation, the counselor's skills in referral and in framing the evaluation request help determine the quality of the resulting evaluation.

Another task of the case manager at this stage is to monitor services as they are delivered. This is important in several respects: for client satisfaction, for the effectiveness of service delivery, and for the development of the case manager's network. Monitoring is doubly important because of the personnel changes that constantly occur in human service agencies. Moreover, there may be a need to revise the plan as problems arise and situations change.

The implementation phase also involves working closely with other professionals, whether they are employees of the same agency or another organization. A case manager who knows how to work successfully with other professionals is in a better position to make referrals that are beneficial to the client. These skills also contribute to effective communication among professionals about policy limitations and procedures that govern service delivery, the development of new services, and expansion of the service delivery network.

Perhaps there is no other point in service delivery at which the need for flexibility is so pronounced. For example, during the implementation stage it often becomes necessary to revise the service plan, which must be regarded as a dynamic document to be changed as necessary to improve service delivery to the client. Changes in the presenting problem or in the client's life circumstances, or the development or discovery of other problems, may make plan modification necessary. Such developments may also call for additional data gathering.

In his second semester at the local college, Roy heard about a course of study that prepared individuals to be interpreters for the deaf. This intrigued him, because he was already proficient in sign language. His mother was severely hearing impaired, and as a child, Roy signed before he talked. He also thought back to the evaluation staff meeting, at which the team discussed the possibility of making interpreter certification a vocational objective for him. Roy liked the interpreting program and the faculty, so he applied to the program. The change in vocational objective made it necessary to modify his plan. His counselor (by now, his fourth) revised the plan at the next annual review to include his new vocational objective of educational interpreting.

 ## Three Components of Case Management

Case review, report writing, documentation, and client participation appear in all three phases of case management; they are discussed in detail in later chapters. Here we introduce the concepts by examining how each applies to Roy's case.

Case review is the periodic examination of a client's case. It may occur in meetings between the case manager and the client, between the case manager and a supervisor, or in an interdisciplinary group of helpers, called a *staffing* or *case conference*. A case review may occur at any point in the case management process, but it is most common whenever an assessment of the case takes place. Case review is an integral part of the accountability structure of an organization; its objective is to ensure effective service delivery to the client and to maintain standards of quality care and case management.

Roy's case was reviewed in several ways. Each time a new counselor assumed the case (unfortunately, this was often), a review was done. There were also reviews on the occasion of the two counselor contacts Roy had per semester. At the end of each semester, his grades were checked—also part of the case review. The staffing about Roy's vocational evaluation is an example of case review by a team. In this case, the client was an active participant in the case review. Roy also participated in developing the service plan, which involved a review of the information gathered, the eligibility criteria, and the setting of objectives. The agency serving Roy implemented the important component of case review in various ways at different times throughout the process.

An important part of case review is the documentation of the case. **Documentation** is the written record of the work with the client, including the initial intake, assessment of information, planning, implementation, evaluation, and termination of the case. It also includes written reports, forms, letters, and other material that furnish additional information and evidence about the client. The particular form of documentation used depends on the nature of the agency, the services offered, the length of the program, and the providers. A **record** is any information relating to a client's case, including history, observations, examinations, diagnoses, consultations, and financial and social information. Also important are "all reports pertaining to a client's care by the provider, reports originating from orders written within the facility for tests completed elsewhere, client instruction sheets, and forms documenting emergency treatment, stabilization, and transfer" (Mitchell, 1991, p. 17). The case manager's professional expertise must include documenting appropriately and in a timely manner and preparing reports and summaries concisely but comprehensively.

Roy's file includes many different types of documentation. In this chapter, the written record includes computer forms, applications for services, counselors' notes, medical evaluations, reports, and letters. Other documentation (not shown here) in Roy's file are a psychological evaluation, a vocational evaluation, specialized medical reports, and medical updates. In Roy's case, all this documentation turned out to be indispensable because he worked with five different counselors. For continuity of service, good case documentation is essential.

Client participation means the client takes an active part in the case management process, thereby making service delivery more responsive to client needs and enhancing its effectiveness. In some cases a partnership is formed between the case manager and the client; an important result of this partnership is client empowerment. One of the many factors involved in forming a partnership with the client is clear communication or two-way communication. The case manager must explain to the client the goals, purposes, and roles of the case manager as defined by the agency. The case manager encourages the client to define his or her goals, priorities, interests, strengths, and desired outcomes. At this point the client also commits to assuming responsibility within the case management process. As client participation continues and the partnership develops, it is helpful for the case manager to have knowledge of subcultures, deviant groups, reference groups, and ethnic minorities so as to communicate effectively with the client about roles and responsibilities. Other factors can affect client involvement, including the timing, setting, and structure of the helping process. Minimizing interruptions, inconveniences, and distractions always enhances client participation.

Encouraging client participation has identifiable components. The first is the initial contact between the client and the case manager. It is easier to involve clients who initiate the contact for help, as Roy did, because they usually have a clearer idea what the problem is and are motivated to do something about it. In Roy's case, the clarification of roles and responsibilities occurred at three points in the assessment phase. Roy and his counselor were able to talk about the agency and the services available, and the counselor encouraged Roy to talk about his goals, motivations, and interests. When Roy completed his application, the counselor went over it with him, especially the statement at the bottom of the second page. On signing the statement, the client voluntarily places himself or herself in the care of the agency. With this agreement come roles and responsibilities for both the client and the counselor, which the counselor reviews at that point. A second opportunity to clarify roles and responsibilities comes with the completion of a service plan. Both the client and the counselor sign the service plan, which designates the responsibility for each task and the time frame for completion of each service.

The middle phase of case management is devoted to identifying problem areas, reaffirming client strengths, and developing and implementing a plan of services. It was during this phase that Roy decided to change his major from business communications to interpreting, which required an amendment to the plan.

Roy also needed help paying for a tutor in a science course that was particularly difficult for him. Client participation during this phase ensures that the client's perceptions of the problem and its potential resolution are taken into account.

The final phase of client participation comes at the termination of the case. At this time, the client and the case manager together review the problem, the goals, the service plan, the delivery of services, and the outcomes. They may also discuss their roles in the process. Thus, in terms of client participation, termination means more than just closing the case. It is an assessment of the client's progress toward self-sufficiency, the ultimate goal of client empowerment. Self-sufficiency is defined differently for each client.

Now that you have some sense of the flow of the process and its component parts, we will review the responsibilities of managing cases in all three phases. This section introduces the principles and goals that guide service delivery and discusses how they influence the work of case management.

 Principles and Goals of Case Management

The guiding principles and goals of case management have emerged through the work of early pioneers in helping, through federal legislation, and through current practice. These include integration of services, continuity of care, equal access to services, quality care, advocacy, working with the whole person, client empowerment, and evaluation. The subsections that follow discuss these principles and their relevance to case managers.

Integration of Services

Integration of services means developing and implementing a plan that brings together a variety of services to help a client. Integrated service delivery is designed so that each service enhances and supports the other efforts and is a guiding principle of case management. Many people enter the system with multiple needs or needs that change over a long period. To address multiple needs, case managers integrate the work of many agencies and professionals. With service integration, there is less chance of fragmentation and duplication. Integrated services also facilitate effective priority setting and encourage positive interaction among the services provided.

> My job is to assist schools with recording their resources and helping them link up with community resources. I have a student health and human service action team, which is where the schools and the community come together. Several major organizations, such as the Board of Education, the Department of Health, the Department of Mental Health, the mayor's office, and the Children's Planning Council, are involved with the structuring of county services to children and families. (Norma Hernandez, Roosevelt High School, Los Angeles, personal communication, March 24, 1998)

Continuity of Care

Continuity of care has two meanings. First, continuity means that services are provided to the client uninterrupted from the first phone call through to termination and often beyond. An individual's needs are addressed, if possible, even before he or she makes the first visit to the agency. For example, the case manager may phone before the initial visit or the intake interview, asking if there are any questions, or he or she might mail a welcoming letter with information about the agency. Continuity of care carries beyond termination, through a transition period when services are no longer needed but the client needs minimal support or just the knowledge that short-term assistance or advice is available if needed.

Second, the term *continuity of care* refers to the comprehensiveness of the care provided. This aspect of continuity involves therapeutic intervention along with support in the environment, maintaining a relationship with the client's family and significant others, crisis intervention, and social networking beyond mere linking of services. This term is also used in reference to managed care, denoting the intensity level of the services provided. For example, a troubled teen might receive in-home counseling, after-school clinical treatment, or commitment to a residential facility. Each represents more intensive intervention.

Jan Cabrera at the Intensive Case Management Program in Los Angeles provides the following illustration of continuity of care.

> The biggest problem is that we have people who come into the hospital to get stabilized on meds; they get a follow-up appointment when discharged and ten days of follow-up medication. They never quite make it to that follow-up appointment and they never quite get the medication refilled. They have two or three weeks worth of meds on board so it takes a while for it to filter out. Eventually they have no more coverage of medication and they become disoriented, psychotic, agitated or whatever. A couple of months later they get called to the attention of the authorities again and they are back in the hospital. (Personal communication, March 24, 1998)

In the situation described, case managers are able to provide a continuity of care that extends beyond hospitalization. As they continue to work with the client, it becomes necessary to closely monitor their follow-up appointments and their medication prescriptions,

Equal Access to Services

Equal access to services means that everyone in need of assistance has the same opportunity to approach, apply for, and use case management services. The commitment to equal access is reflected by the case manager assuming the role of advocate. Attention is also given to developing ways to extend access to services, such as fee waivers, transportation, and outreach efforts. Affordability is directly linked to the issues of eligibility and access. To ensure access to services, eligibility must be defined so as to include those who lack traditional economic, social, and political access. There must also be a plan for individuals who are on the margins of eligibility. Walter Rae, a case manager at a Houston hospital, discusses a problem they encounter.

> An individual recently came to the emergency room because of a suicide attempt. He was also a cross-dresser, going to become a female. He also had full-blown AIDS. He probably had about six or seven months to live. Very narcissistic. Complete denial that he was going to die. And a lot of what had to occur was getting him into a place that would accept him, like shelter, because they don't like to take people who are that odd, even in this area, in an all-men facility. It was a very difficult placement. (Personal communication, December 16, 1999)

Quality Care

Quality care is providing superior services to all clients and implies a commitment that respects the rights of the client and demands accountability on the part of human service professionals. The watchwords are effective and efficient care; both are key considerations in service delivery. Quality assurance is a component of case management; the term signifies professional excellence, high standards of care, and continuous improvement (Mullahy, 1998).

Workers at Youth in Need in St. Louis point out that success is defined in different ways, depending on the client and the goals of the service.

> With outreach we have different levels of success. For some people that we meet, success is getting them off the streets and into a shelter. For the ones that don't want to get off the streets, for example, those that don't want to go to a shelter, they don't want to go somewhere else to stay. Maybe they just want us to help them get through the winter or get them food, or warm clothes. These are different cases because they are not trying to get the same thing. Some of them are just trying to get food to survive while others want to get to a place that is warm. It is different level of success. (Rick Langley, Youth in Need, St. Louis, personal communication, December 13, 1999)

Effectiveness means getting results. In this age of scarce resources, it is important that the available resources are used wisely, that is, *efficiently*. In service delivery, the productive use of resources involves determining the outcomes desired and developing a plan to achieve those outcomes. Efficiency is measured in terms of the resources required, the time expended, the cost of services, and the outcomes achieved. The plan is constantly monitored, necessary adjustments are made, and appropriate justification is presented when additional resources are needed.

The case manager must maintain efficiency because of the complexity of many of the problems he or she faces. It is essential that professionals work together to deliver quality services in an efficient manner.

Advocacy

Advocacy within the case management process is the act or process of representing the interests of the client and teaching the client to advocate for him- or herself. It may involve speaking or writing in defense of a person or cause. In representing the interests of the clients, case managers often assume a dual role, representing both the institution and the client. Advocacy is essential because of imperfections in the delivery system: duplication, fragmentation, and resources that are often insufficient to provide the services needed. Sometimes the needed services do not exist. Of course, there must be adequate resources to manage in order to provide quality care.

Linda Washington and her colleagues at the Third Avenue Family Service Center in the Bronx describe their advocacy work:

> When we go into the home we also advocate services. . . . A homemaker may need something that has not been done. Appointments have not been kept, so we need to reschedule new appointments. . . . Transportation is needed. . . . Services have not been provided at a speedy rate; so we again intervene. (Personal communication, May 5, 1994)

The Whole Person

Case managers must be committed to a holistic view of the individual receiving help. This means they recognize that they work with the **whole person**, acknowledging the many human dimensions to be considered in service delivery: social, psychological, medical, financial, educational, and vocational. Most likely, the client has problems in more than one of these areas.

One family counseling agency uses a 12-page standardized form during the intake interview. The Pima Client Assessment Plan is designed to uncover a range of problems clients may have.

> It is a review of their medical functioning, social support system, needs for medical equipment. . . . This form is an outline. . . . It is the ticket to services within our system. . . . We also look for more than just what is on the form. . . . We are looking for a picture of their lives. (Suzy Bourque, Family Counseling Center, Tucson, AZ, personal communication, October 7, 1994)

Client Empowerment

Client empowerment within the case management process means respecting clients as individuals, building on their strengths and interests, placing them in a partnership role, and moving them toward self-sufficiency. Respect for the client stems from the long-standing belief that all individuals, regardless of their needs or disabilities, have integrity and worth. This belief guides the case manager to place the client in a central role in the helping process. Ensuring the client's full participation during the service delivery process means involving him or her in every step, including identifying the problem, gathering information, establishing goals, planning, implementing the plan, and evaluating the plan and outcomes.

Client empowerment within the process expands respect for the client to include teaching basic service coordination skills and, as time and skill development allow, encouraging the client to manage his or her own case. The goal is to develop self-sufficiency so that the client can manage his or her own life without depending on the human service delivery system.

In many agencies, self-sufficiency is a primary goal, as Janelle Stueck says.

[We are advocates] for the participant . . . trying to help that person become self-sufficient. We try to help them make plans so they will know from step 1 where they need to go, how they need to go about it, and what the end result will be. . . . The person may come in very dependent . . . but hopefully, as things progress, they become more sure of themselves . . . and they are able to do more and more on their own. And so hopefully by the end they are quite ready to leave the nest, as it were. (Personal communication, August 30, 1993)

Another way of showing respect for the client is to treat him or her as a customer. This means asking clients about their needs, providing services that match those needs, offering clients an opportunity to evaluate the services they receive, and making changes to improve services based on their feedback. Sick Kids Need Involved People (SKIP) of New York is an agency that has worked hard to treat its clients as customers. It was a struggle for the agency to make this transition.

To empower the families, a lot of emphasis is placed upon parent partnerships. Parents or caregivers should be provided with information and choices so, as knowledgeable participants, they can make informed choices and decisions on behalf of their children. Professional input is often riddled with highfalutin terminology and statements of "best care" for the child. Parents need to understand professional input; parents deserve the respect of all parties involved with their child. As [customers], they seek the best and often require guidance, information, and training to realize the goals they have designed for the child. This agency was founded by the parents of a child with complex medical needs. Based upon this genesis, the organization has evolved and succeeded based upon parent [customer] feedback, involvement, and partnership. (Margaret Mikol, SKIP, personal communication, May 4, 1994)

Client empowerment also means providing services based on client strengths. This focus begins as early as the initial assessment phase, when the case manager and the client begin to identify positive client characteristics on which to build a treatment plan.

Evaluation

As a critical part of case management, *evaluation* is the assessment of the process, the outcomes, and the quality of the process. Evaluation takes place throughout the case management process. The focus is on relevance for the client, client progress and satisfaction, integration of services, quality of services, and outcomes. Both professionals and clients are involved in the evaluation process.

> They review the program every 6 months unless we see a change in behavior and we think that something needs changing . . . or there is no need for this program . . . or someone is told this medication is not working. (Kim Ehlers, Mitchell Area Adjustment Training Center, Parkston, SD, personal communication, October 14, 1995)

Chapter Summary

Managing client services is an exciting and challenging responsibility for helping professionals who work as case managers. To assist clients with multiple problems, case managers must know the process of case management and be able to use it. The process can be adapted to many different settings, for work with a variety of populations. The coordination and the provision of direct services are guided by common principles and goals, committing the agency to give the client quality services and ensuring that he or she participates in the case management process.

The three phases of case management—assessment, planning, and implementation—each represent specific responsibilities assumed by the case manager. The process of case management is nonlinear, for example, a case manager may make assessments early on and return to conduct assessment during the planning and implementation work with the client. Three components of the case management process appear in all three phases of case management: case review, report writing, and documentation. Note that each component includes intense interaction with and participation by the client. These components require ongoing evaluation and written documentation of the case activities. The process of case management described here is based on important principles and goals of case management. These include integration of services, continuity of care, quality care, advocacy, working with the whole person, client empowerment, and evaluation. The purpose of these principles and goals are to provide the most effective and efficient services to the client.

Chapter Review

◈ *Key Terms* ....

Case management	Integration of services
Case manager	Continuity of care
Assessment	Equal access to services
Planning	Quality care
Implementation	Advocacy
Case review	Whole person
Documentation	Client empowerment
Record	

◈ *Reviewing the Chapter* ....

1. Describe the context in which case management services are delivered today.
2. How has the definition of case management changed over the years?
3. What is the expanded definition of case management?
4. Distinguish between the terms *applicant* and *client.*
5. What should be accomplished during the assessment phase?
6. What occurs during the initial contact between the case manager and the individual seeking services?
7. Describe the routine matters that are discussed during the initial contact.
8. Identify the types of information that are gathered during the initial interview.
9. Which factors determine what information has been obtained by the end of the initial interview?
10. Using the case presented in this chapter, discuss the advantages of a partnership between the case manager and the client.
11. Describe the case manager's activities during the planning phase.
12. What questions guide implementation?
13. Why is flexibility so important during the implementation phase?
14. Define *case review.*
15. List the three keys to successful case review.
16. Why is documentation important in service coordination?
17. How can the case manager promote client participation?
18. How will a client's resistance affect his or her participation in the service coordination process?

◈ *Questions for Discussion* ....

1. Why do you think that case management as a method of service delivery is changing?
2. From your own work and study of human services, what evidence do you have of the importance of assessment and planning?

3. If you were a case manager, what three principles would guide your work? Provide a rationale for your choices.
4. What do you think Roy Roger Johnson would say about his experience with the case management process?

References

Jackson, B., Finkler, D., & Robinson, C. (1992). A case management system for infants with chronic illnesses and developmental disabilities. *Children's Health Care, 21*(4), 224–232.

McClam, T., & Woodside, M. (1994). The practitioner's voice: Case management for effective service delivery. *Human Service Education, 14*(1), 39–45.

Mitchell, R. W. (1991). *Documentation in counseling records*. Alexandria, VA: American Association for Counseling and Development.

Mullahy, C. M. (1998). *The case manager's handbook*. Huntington, NY: Aspen.

Sullivan, W., Wolk, J., & Hartmann, D. (1992). Case management in alcohol and drug treatment: Improving client outcome. *Families in Society: The Journal of Contemporary Human Services, 73*(9), 195–204.

Webster's New Collegiate Dictionary (1st ed.). (1973). Boston: G. & C. Merriam.

Historical Perspectives on Case Management

A *lot of families fall through the cracks. . . . We come in and we do casework for them . . . we provide financial assistance . . . we assist with telephone bills . . . we go into the home and advocate services.*
> —Carolyn Brown, Third Avenue Family Service Center, Bronx, NY, personal communication, May 5, 1994

I *t is not competitive between the agencies. . . . We have case managers who are gatekeepers of the system. . . . We have the central and northwest part of the city. . . . Most of our clients come from a hospital discharge planner.*
> —Suzy Bourque, Family Counseling Center, Tucson, AZ, personal communication, October 7, 1994

I *n one sense, we are not really state employees; but in another sense, we are. I think that it is a new trend in government where agencies are privatized so they can mainstream service delivery to youth. The office developed as a result of the governor's order and we are part of an agency that does a variety of things, but we are now under the Department of Health.*
> —Thom Prassa, East Tennessee Community Health Agency, Knoxville, TN, personal communication, July 10, 1994

The purpose of this chapter is to establish a historical context for case management. It will describe four perspectives on case management that have evolved in the past 30 years: case management as a process, client involvement, the role of the helper, and utilization review and cost–benefit analysis. There follows a brief history of case management in the United States, including its evolution to broader service coordination responsibilities within the last decade, as well as an introduction to managed care.

By the end of each section of the chapter, you should be able to accomplish the performance objectives listed below for that section.

PERSPECTIVES ON CASE MANAGEMENT
• Identify four perspectives on case management.
• Trace the evolution of case management.
• Describe the impact of managed care organizations on case management and service delivery.

THE HISTORY OF CASE MANAGEMENT
• Assess the contributions of the pioneers in the areas of advocacy, data gathering, recordkeeping, and cooperation.

- Using the Red Cross as an example, describe casework during World Wars I and II.
- Name the acts of federal legislation that further developed case management.

THE IMPACT OF MANAGED CARE
- List the goals of managed care.
- Summarize the impact of managed care on human service delivery.
- Differentiate between the types of managed care organizations: HMOs, PPOs, and POS.

EXPANDING THE RESPONSIBILITIES OF CASE MANAGEMENT
- Trace the shift in emphasis in case management.
- Explain the strengths and weaknesses of managed care.

Case management has long been used to serve human service clients. Today, professionals are discovering new and more effective ways to deliver services, and there is no longer a standard definition. Modern-day case management does resemble the practice of the past, but many dramatic changes have occurred. Among them are the changing needs of individuals served; financial constraints on the human service delivery system; the increasing number of people needing services; and the growing emphasis on client empowerment, evaluation of quality, and service coordination.

One consistent theme that runs through the study of human service delivery is diversity. The three helping professionals quoted at the beginning of this chapter describe the services of their agencies. The case worker from the Third Avenue Family Service Center describes the work of her agency as providing financial assistance and advocacy work for families for whom there is no other support. The Family Counseling Center in Tucson, on the other hand, gives patients just discharged from the hospital only the aftercare services they need. This often includes support to meet psychological, social, medical, financial, and daily living needs. In Thom Prassa's case, services to youth have changed recently. The state contracts with the agency where he works to deliver services. These services, like those of the other organizations, are delivered within the context of federal and state regulations and restrictions.

Much of the foundation of case management developed when it was used to serve people with mental illness who were deinstitutionalized in the 1970s. Illustrating our discussion here is the story of Sam, who was diagnosed as mentally ill and promptly institutionalized. Sam has received many services since that first diagnosis, and his history with the human service delivery system reflects the evolution of service delivery from the traditional form of case management to the new paradigm that is applied today.

 Perspectives on Case Management

This section explores four different perspectives on case management, which together serve to illustrate the development of case management since the 1970s.

Case Management as a Process

In the 1970s, the mental health community was involved in the process of **deinstitutionalization**: the movement of large numbers of people from self-contained institutions to community-based settings, such as halfway houses, family homes, group homes, and single residential dwellings. The following definition of case management was offered by a member of the American Psychiatric Association's Ad Hoc Committee on the Chronic Mental Patient.

> My view is that it is a vital, perhaps the most primary, device in management for any individual with a disability where the requirements demand differential access to and use of various resources. Far from a new concept, it has long been the central device in every organized arrangement that heals, rehabilitates, cares for, or seeks change for persons with social, physical or mental deficits. . . . Case management is a key element in any approach to service integration. . . . A counselor manages assessment, diagnosis, and prescription . . . synthesizes information, emerges with a . . . treatment plan; and then purchases one or more interventions. (Lourie, 1978, p. 159)

Many clients need assistance in gaining access to human services. Often they have multiple needs, limited knowledge of the system, and few skills to arrange services. Sam's case reflects the experiences of many clients who were institutionalized in the 1950s and 1960s and were later deemed appropriate for discharge during deinstitutionalization. The case management process illustrated next is an elementary one: limited assessment followed by placement.

After several ear infections, Sam became severely hearing impaired when he was 3. He was the youngest of four children and lived in a small town with his mother, two sisters, and a brother. His mother took care of him, and he became dependent on her. They learned to communicate with each other using a sign language they devised themselves. None of his siblings learned to sign. Sam was often unruly and found that tantrums would get him what he wanted. The older he got, the harder to handle he became. When Sam was 15, his mother died. None of his siblings would assume responsibility for him, so they decided to have him admitted to the state mental hospital in the capital. This occurred in the early 1950s; the exact date is unknown because a fire at the institution in the 1960s destroyed the records of those who were admitted previously. Sam was in the mental institution for many years before the deinstitutionalization movement started. At that point, Sam's long odyssey began. (See Table 2.1.)

Sam's first case manager was an employee of the institution, and his job was to identify patients who could function in a community setting. Limited assessments of Sam's mental and emotional state indicated that he was not mentally ill but simply hearing impaired. Unfortunately, his time in the institution had compounded his problems: He did not know American Sign Language

TABLE 2.1 SAM'S JOURNEY

Sam's age	Provider	Activity
3	Mother	Care at home
15	Institution	Residential care
26	First case manager	Limited assessment
	Sister	Lived with sister for one weekend
	Group home	Lived in group home for six days
	Institution again	Residential care
27	Second case manager	Assessment
	Third case manager	Learning American Sign Language
		Contact with deaf community
		Planning, involving Sam in process
28		Independent living skills
29	Day care program	Socialization skills
30		Fluent in American Sign Language
		Developed friendships
30–40	Lois Abernathy (care coordinator)	Regular visits and assessment
	Interdisciplinary team	3 halfway houses; 4 group homes;
		School for the day
		Living with siblings
		Living in apartment
		Vocational training
40	Rehabilitation counselor and	Vocational training
	Care coordinator	Living with friend in apartment
41	Rehabilitation counselor	Full job responsibility
		Termination of services

(ASL), and he had begun to behave like other patients who did have mental illness. The case manager decided that Sam should be moved from the institution to another setting. Sam's oldest sister was located, but he stayed with her for only one weekend. She returned Sam to the institution on Monday morning, saying that she couldn't handle him and his presence was too disruptive to her family. None of his other siblings was willing to help, so Sam remained in the institution while his case manager searched for a group home that had an opening. Eventually, Sam did move into a group home, but he lived there for

*just six days before returning to the institution. According to the home's direc-
tor, no one could communicate with Sam, his behavior was inappropriate, and
he needed constant supervision.*

The responsibility of Sam's case manager was to find Sam an environment
that could foster his growth and development. Unfortunately, the search for such
an environment was quite difficult, given Sam's institutional behavior and the
limited assessment of his abilities. Sam is one of those clients who need access to
multiple services before they can make the transition from an institution to a
community setting.

One way to think about the case management process is the key elements for
success: responsibility, continuity, and accountability (Ozarin, 1978). *Responsibility* means that one person or team assesses the client's problem and then plans
accordingly. Linkages "must be established to form a network of service agencies
which can provide specific resources when called upon without assuming total
responsibility for the client, unless responsibility for carrying out the total plan is
also transferred and accepted" (p. 167). In other words, there must be a clear line
of responsibility for the case and the client.

Continuity is another significant element of good case management. Planning
is the key that ensures continuity. It is important not only during the intensive
treatment phase but also in aftercare. To foster *accountability*, methods "must be
in place to assure the patient is not lost. . . . The case management process must
help the client increase the ability to function independently and to assume self-
responsibility. The client should be involved in all aspects of decision making"
(Ozarin, 1978, p. 168). Guided by these goals, organizations and professionals at
every level work hard to develop systems that participants understand, working
together to serve and involve the clients.

Case management can be seen as a "set of logical steps and a process of interaction within a service network which assure that a client receives needed services in a supportive, effective, efficient, and cost-effective manner" (Weil & Karls,
1985, p. 2). Case management is an important and necessary component of the
human service delivery system, for it provides a focus and oversees the delivery
of services in an orderly fashion. Weil and Karls agree with Lourie (1978) that
case management is needed because of the multiple needs of clients and the
complexity of the delivery system. By the early 1980s, a model of intensive case
management was introduced. One example of the model was Assertive Community Treatment (ACT) adapted from an earlier Training in Community Living
(TCL) program developed in Madison, WI. The ACT programs offered to mentally ill clients were intensive case management services while living in the community. These programs established quality standards of care for clients by
emphasizing assertive outreach as well as care giving (Kuno, Rothbard, & Sands,
1999). As you read about Sam's case, you will see case management evolve into
a more logical and complex process that focuses on client participation, integration of services, and cost effectiveness.

Client Involvement

In the 1980s, client involvement came to be emphasized more strongly. A model of case management was proposed based on the concept of enabling clients "to solve problems, meet needs, or achieve aspirations by promoting acquisition of competencies that support and strengthen functioning in a way that permits a greater sense of individual or group control over its developmental course" (Dunst & Trivette, 1989, p. 93). Sam's experience in the human service delivery system reflects the beginning of changes in service provision.

Sam remained institutionalized for the next four months because there was a shift in the case management process. The institution decided to contract with a local mental health agency for case management services. Case management was a new role and responsibility for this agency; it assigned two individuals 50 cases each, with few guidelines for performing this new function. Sam's new case manager, his second, spent some time assessing Sam's needs, getting to know him, and talking with the mental health professionals within the institution. In concert, they determined that Sam needed a very structured environment in the community if his deinstitutionalization was to succeed. He also needed to learn sign language and to begin to communicate with others using this medium. Because he was hearing impaired but did not have mental illness, he needed to be in contact with the deaf community, where he could find support and role models for independent living. At the age of 27, he had little ability to care for himself.

Unfortunately, his case manager left her position before she had the opportunity to implement the plan. A third case manager assumed responsibility for Sam's case, with much determination. Her own brother had been hearing impaired since birth, and she recognized Sam's potential. She could communicate with him using ASL. She was also committed to planning, documenting her work, involving Sam in the case management process, and following through on referrals and the involvement of other professionals.

The goals of the plan included having Sam learn ASL, teaching socialization skills and independent living skills, and introducing him to members of the deaf community. His case manager was able to find a day care program where Sam could learn independent living skills. Three times a week, he went to the local School for the Deaf for ASL lessons. Once a week, Sam and his case manager joined other hearing impaired adults for a special community program and social hour. Sam still lived at the institution. By the end of the second year, the case manager was able to include Sam in the process of setting priorities and planning for his treatment.

After two years, Sam had made considerable progress. He was able to use ASL to communicate his needs, and he had developed several friendships with people he met at the School for the Deaf and at the community programs. On his

30th birthday, he celebrated with his friends from the school. Tantrums continued to occur, but less frequently.

After a rocky beginning, Sam benefited from the service delivery process as it evolved. His third case manager assumed responsibility for his case, provided the continuity needed for him to make progress, and was accountable for his care. She used a process of logical steps to establish goals and set priorities. She also established a partnership with Sam by involving him in problem identification, plan development, and service provision. By learning daily living skills, Sam reinforced his ability to care for himself, and his increasing mastery of ASL gave him a new medium for self-expression and communication.

The Role of the Case Manager

Chapter 1 lists an array of job titles that have emerged to reflect the new goals of service delivery. Traditionally, terms such as *caseworker* and *case manager* described the efforts of helpers. Today, job titles include *service coordinator, liaison worker, counselor,* and *case coordinator*. These new job titles represent not only the diversity of service delivery today but also the broader range of responsibilities and the different ways case managers perceive their roles. The change in job titles reflects the evolution of case management and, in a larger context, of service delivery. The emphasis shifted from what was previously understood to be case management, when it meant the skills of managing someone, to terminology reflecting a more equitable relationship, such as *coordination* and *liaison*. A change in philosophy had occurred regarding the role of the case manager, emphasizing working with other professionals, coordinating care and other services, and empowering individuals to use the system to help themselves. The focus became the client's ability to develop the skills needed to work within the human service network.

The next 10 years were a struggle for Sam and for those who worked with him. His case manager of two years left her job for a promotion in a nearby city, and his case was transferred to Lois Abernathy, a care coordinator at a different agency. Because of increasing pressure to deinstitutionalize, it was decided to move Sam to a halfway house before helping him establish residence in the local community. Over the course of the decade, Sam lived in three halfway houses, four group homes, the School for the Deaf, with his siblings, and in an apartment with a roommate. Ms. Abernathy was the link between Sam and each of these placements. Her responsibilities included meeting with Sam regularly to review his needs, problems, and successes and arranging any additional services for him. Often, she and Sam would meet with other professionals who were involved with his case. Ms. Abernathy was committed to giving Sam choices about his future. When he expressed the desire to work at a job, she helped him determine exactly what he would like to do. After exploring the options available to him, Sam decided that he would like to work with a local business vending program that Ms. Abernathy knew about. After

> *much education and training, his responsibilities with this program came to include stocking machines, collecting money, and minor repair work.*

The assignment of a *care coordinator* to Sam's case signaled a shift in the role of the case manager—from management to coordination. The client's participation in the process also became significant; a partnership emerged to identify, locate, link, and monitor needed services. In this way, case management built on Sam's strengths and empowered him to help himself.

Utilization Review and Cost–Benefit Analysis

One result of the spiraling cost of medical and mental health services and the push for health care reform is the growth of the managed care industry. The purpose of managed care is to authorize the type of service and the length of time care is provided and to monitor the quality of care. In the managed care environment, case managers function very differently from those described earlier.

What makes case management in managed care distinctive is the emphasis on efficient use of resources. Case managers are involved in utilization review and have the responsibility to authorize or deny services. Also, they must know how to interact with insurance providers and how to process claims through the insurance system.

This new case manager is also responsible for cost–benefit analysis. Such an analysis does not include the traditional reporting that is found in a case history as notes or recommendations and follow-up. It is focused on the financial matters of the case, specifically the cost and efficiency of services.

> *As we leave Sam at the age of 42, he is two months away from assuming responsibility for all the vending machines in a nearby neighborhood. It has taken two successive six-month training sessions to teach Sam the necessary job skills and repair techniques. Sam is living with a new friend in a small apartment near his vending area. This friend is also a client at the rehab center and is helping Sam train for his new job. Sam's rehabilitation counselor and his care coordinator from the mental health center have met and developed a coordinated plan, with input from Sam. Because of his recent success in rehabilitation, mental health services are no longer authorized for Sam.*

Sam's experience with a care coordinator has led him in a new direction. The emphasis on tapping Sam's potential and coordinating care has given him a major voice in decision making. This requires coordination between the two systems of rehabilitation and mental health. Sam does make progress, and the managed care case manager decides to discontinue mental health services. As we leave Sam, his rehabilitation counselor supports him job training and housing. If he should again need professional mental health support, it is hoped that the rehabilitation counselor can arrange these services for him.

The History of Case Management

As important as various current perspectives on case management are, the historical roots are equally informative. The pages that follow trace the history of case management from its origins in institutional settings through the work of early pioneers, the impact of the American Red Cross, and the influence of federal legislation. The chapter concludes with a discussion of case management as it is practiced today.

"The process of service coordination and accountability has a century-long history in the United States" (Weil & Karls, 1985a, p. 1). As first used in institutional settings, case management included the responsibilities of intake, assessment of needs, and assignment of living space. These institutions provided residential services to people incarcerated for crimes, orphans, people with mental illness, people with disabilities, and elderly people. Which professionals performed the case management function depended on the particular institution; among them were doctors, nurses, psychiatrists, psychologists, counselors, and teachers (Weil & Karls, 1985b).

Early Organizations

One example of an institution with an early commitment to case management was the Massachusetts School of Idiotic and Feebleminded Youth, established in 1848. This school promoted the belief that people categorized as "idiotic" or "feebleminded" could improve if they were given appropriate clinical, social, and vocational services and support (Weil & Karls, 1985b). In 1839, a child who had mental retardation as well as vision impairment was brought to the Institution for the Blind. It was clear that the child had needs beyond the expertise of the institution, and the director, Samuel Howe, was determined to help this child and others with similar needs. He convinced the state that he could improve these children in three areas: bodily habits, mental capacities, and spirituality (Winsor, 1881).

The new institution founded under Howe's leadership was the Massachusetts School of Idiotic and Feebleminded Youth. It provided pioneer services in case management, although the school did not use that term. Early services at the school included observation and diagnosis of physical and mental behavior. The helping professionals tracked clients' progress, and they soon began differentiating and individualizing treatment: ". . . we cannot properly care for a young and helpless idiot in the houses devoted to the brighter moron children" (Trustees, 1920, p. 17).

When demand for the services increased, the school established outpatient clinics in several cities (Trustees, 1919). These clinics supported families who cared for children at home. Aftercare was also an important service, provided by trained "visitors" who helped plan the transition from the institution to the home or other setting. The visitor would gather information to determine whether the child should be sent home on trial or be released for vacation with family or friends. They would also follow up after transition to determine whether the release and placement were appropriate. It was assumed that aftercare (supported by parents, family, and the visitors) could "save much money to the state by helping to continue the custody and training of the troublesome defects, and to permit the liberty of those who can safely use such liberty" (Trustees, 1920, p. 19). This emphasis on aftercare was the forerunner of modern-day continuity of care, as well as today's commitment to provide services as the client makes a transition from the treatment setting to a less restrictive one.

Early in this century, the school made two improvements in the management of information, which is an important component of case management. In 1916, the institution began an evaluation of its services. This included a study of clients who had been discharged and those in aftercare. The information gathered included where patients lived, with whom they were living, whether they supported themselves, and (if so) how. The information was gathered by survey, and then patients, families, and friends were interviewed in their homes. In 1919, new legislation established a Registry for the Feebleminded in an effort to catalog the state's population of people with retardation (Trustees, 1920). These advances in recordkeeping and information management contributed to case management as we know it today.

In the 1860s, the state of Massachusetts created a Board of Charities. The board had responsibility to coordinate human services and to monitor public expenditures for the poor and the sick (Weil & Karls, 1985b). In addition, Charity Organization Societies were established in New England to provide service coordination and interagency cooperation. Their primary responsibility was to register clients and needy families, and an effort was made to reduce duplication of services. The Charity Organization Societies' friendly visitors "had to be trained in investigation, diagnosis, preparation of case records, and treatment, all of which required guidance, counsel, supervision, and the knowledge of the scientific philanthropy" (Trattner, 1994, p. 238).

The next section discusses the work of some early pioneers in human services who developed case management further, especially in coordination of services and interagency cooperation.

Early Pioneers

Early case management took either of two forms: a multi-service center approach or a coordinated effort of service delivery. Jane Addams, Lillian Wald, and Mary Richmond were three early pioneers who contributed to the development of the emerging case management process.

HULL HOUSE
Hull House was founded in Chicago in 1889 by Jane Addams and Ellen Starr, classmates at the Rockford Female Seminary. In their twenties, while traveling in England, they visited Toynbee Hall, a university club that had established a recreation club for the poor. Addams and Starr were committed to increased commu-

nication between the classes, and the activities of Toynbee Hall inspired them. They returned to America and moved into a poor section of Chicago, hoping to improve that environment (Addams, 1910).

They bought Hull House, an older home on Chicago's West Side. Committed to sharing their home and their love of learning, they opened the house to the neighborhood. They acquired collections of furniture, art, and literature; soon they became involved with music and crafts. The purposes of Hull House were threefold: to provide a center for civic and social life, to improve conditions in the neighborhood, and to provide support for reform movements (Addams, 1910).

At first, the activities of Hull House involved only casual interactions with the neighborhood, but soon the bonds became more formal. Clubs enrolled members, and regular classes were held. This increased the need for effective administration and recordkeeping. Hull House was intended to be a gathering place that fostered a sense of community and friendship, so many participants would have been suspicious of formal files being kept, attendance logged, records of meetings maintained, or notes written about interactions or behavior. However, neighborhood residents began using the staff members as references, and more detailed records had to be kept. Then information about households was gathered and recorded, just to maintain successful contact. This demographic information was used to generate mailing lists for announcements of classes and other activities. A card system was housed in the administrative office, and its contents were shown only to people who could establish a need to see it (Woods & Kennedy, 1911).

In addition to recordkeeping, advocacy was a case management function that was integrated into the work of Hull House. Jane Addams and her colleagues were involved in many efforts to improve the living and working conditions of the neighborhood and its inhabitants. In one such project, they worked with landlords and lawyers to improve existing housing and find better housing. First, they collected data about the types and characteristics of housing in the neighborhood. In order to understand the purchasing power of the renters, they also studied wage levels (Residents, 1985). Since the residents themselves served as the data gatherers, they became experts on the conditions of their environment (Brieland, 1990). Advocacy took many forms at Hull House; at one time, Jane Addams helped correct unsanitary conditions by becoming a health inspector for garbage collection (Polacheck, 1989). (See Box 2.1.)

HENRY STREET SETTLEMENT HOUSE

The Henry Street Settlement House organization was established in 1895 in New York by Lillian Wald and Mary Brewster, who were nurses. Early in their careers, they decided to provide services to New York City's Lower East Side, home to a large number of immigrants. Wald and Brewster lived in the neighborhood and provided health care services.

They established a system for nursing the sick in their own homes, promoting the dignity and independence of the patient. According to Wald (1915), "the nurse should be as ready to respond to calls from the people themselves as to calls

BOX 2.1   **A Hull-House Girl**
● ● ● ● ● ● ● ● ● ● ● ●

Hilda Satt Polacheck was born in 1882 in Wolclawek, Poland. She was the eighth of twelve children born to a well-to-do Jewish family. Because of oppression by the Russian government, her family immigrated to America in 1892. During their early years, Hilda and her sister attended the Jewish Training School on Chicago's West Side. After the death of their father, however, they joined the ranks of the working poor. Hilda began working in a knitting factory when she turned 13. She first came to Hull House for a Christmas party in 1896, and it soon became the center of her social life. She attended classes and club meetings, read literature, exercised, and performed in plays there. Later she worked there as a receptionist and a guide. Hilda grew up with Hull House and spent much of her time there from 1895 until 1912. In her autobiography, she provides the following description of Jane Addams and her advocacy work.

> Bad housing of the thousands of immigrants who lived near Hull-House was the concern of Jane Addams. Where there were alleys in the back of the houses, these alleys were filled with large wooden boxes where garbage and horse manure were dumped. . . . When Jane Addams called to the attention of the health department the unsanitary conditions, she was told that the city had contracted to have the garbage collected, there was nothing it could do. . . . She was appointed garbage inspector for the ward. I have a vision of Jane Addams . . . following garbage trucks in her long skirt and immaculate white blouse. . . .
>
> Hull-House was in the Nineteenth Ward of Chicago. The people of the Hull-House were astonished to find that while the ward had 1/36 of the population of the city, it registered 1/6 of the deaths from typhoid fever. Miss Addams and Dr. Alice Hamilton launched an investigation that has become history in the health conditions of Chicago. . . . Whatever the causes of the epidemic, that investigation, emanating from Hull-House, brought about the knowledge of the sanitary conditions of the Nineteenth Ward and brought about the changes we enjoy today.

SOURCE: From *I Came a Stranger: The Story of a Hull-House Girl*, by H. Polacheck, pp. 71–72. Copyright © 1989 University of Illinois Press. Reprinted with permission.

from physicians, that she should accept calls from all physicians, and with no more red tape or formality than if she were to remain with one patient continuously" (p. 27). There was an explicit focus on accessibility.

The Henry Street Settlement House kept careful records of its services and made significant contributions to knowledge about the prevalence of disease, as well as treatment. For example, in 1914, the staff at Henry House cared for 3,535 cases of pneumonia, and there was a mortality rate of 8.05% (Wald, 1915). This recordkeeping reinforced the idea that maintaining and managing information was an important part of the process of delivering services. The settlement house was committed to integrating its services with those of the school and the home. The staff kept data on schoolchildren and identified children who had traditionally been excluded from schools for medical reasons. The staff then served as advocates for educational services for these children.

The work at Henry Street led to two significant innovations: the designation of the visiting nurse and the development of the Red Cross. Both these services were important in promoting public health. One important function of the visiting nurse was to establish an organized system of care and instruction for people with tuberculosis and their families. Early in its history, the Red Cross facilitated the use of the public schools as recreation centers, taught housekeeping skills to women, and provided penny lunches for children (Wald, 1915).

Much of the work of Jane Addams and Lillian Wald was conducted with immigrant populations in urban areas. Mary Richmond, a social reformer at the turn of the century, had a similar commitment to bettering the lives of this population. She, too, made significant contributions to the development of case management. Richmond promoted the idea that each person was a unique individual whose personality, family, and environment should be respected. Working with immigrants, she emphasized the need for social workers to resist the tendency to stereotype or overgeneralize (Lieberman, 1990). Richmond wrote, "the social adjustment cannot succeed without sympathetic understanding of the old world backgrounds from which the client came" (1917, p. 117). She also believed that professionals should work with clients rather than doing things to them.

One method Richmond developed to focus on the individual was **social diagnosis**, a systematic way for helping professionals to gather information and study client problems. She established a series of methods for gathering information about individuals, assessing their needs, and determining treatment. This process is often referred to as social casework, and it became a part of the case management process. Richmond contributed a case record form designed to focus on individuals and their unique problems (Pittman-Munke, 1985; Trattner, 1994).

She also recognized that gathering data is a complex process and urged the use of different methods for different individuals. According to Richmond (1917), some clients should be interviewed in an office; for others, the home was the preferred location. She believed in multiple sources of information and warned that data gathering was a complex and often incomplete process.

Cooperation was a continuing goal for Richmond in developing the theory of social casework. This emphasis on coordinated care extended to her work with individuals, and she later expanded that work to include families as well (Montalvo, 1982). Richmond also fostered coordination among agencies. For example, her work in Philadelphia with the Charity Organization Society facilitated relationships between that society, the board of education, and the public education association (Pittman-Munke, 1985).

During the early part of the 20th century, the Red Cross emerged to meet the multiple needs of individuals. The subsection that follows describes its involvement in providing assistance, primarily to servicemen and their families in World War I and World War II.

The Impact of World Wars I and II and the American Red Cross

During the First World War, there was an increased interest in casework as developed by Mary Richmond. The American Red Cross, whose roots are found in the work of Clara Barton and the Civil War, used casework to address individuals' problems and their psychological needs. The use of a casework approach to assist individuals began during the Mexican civil war (1911–1917), when the American Red Cross provided a variety of services to support the daily life of civilians and troops along the Mexican border (Dulles, 1950). Subsequently, the services were extended to dependents of military personnel in army installations throughout the country. These services, performed by the Home Corps of the Red Cross, helped address the needs of the families of military personnel.

In World War I, the Home Service Corps (later known as the Social Welfare Aide Service) provided help to families experiencing problems such as illness and marital difficulties. During the war, the Home Service Corps spent approximately $6 million and handled 2 million cases. During the Second World War, this work expanded: Expenditures climbed to $55 million, and the number of cases handled increased to 10 million (Hurd, 1959). The key was coordination with other welfare agencies, and the Red Cross entered into a formal agreement of cooperation with the Family Welfare Association of America.

The Home Service Corps faced a wide variety of problems. Dulles (1950, pp. 391–392) presents the following example of messages received and sent by the Home Service Corps.

Incoming: MILITARY AUTHORIZE INFORM FAMILY SERVICEMAN WELL ON ACTIVE DUTY NOT REPEAT NOT THE MAN THEY SAW IN NEWSREEL

Outgoing: MESSAGE DELIVERED FAMILY MUCH RELIEVED MOTHER IMPROVING

Incoming: SERVICEMAN REQUESTS MATERNITY REPORT WIFE EXPECTING CONFINEMENT EARLY JULY

Outgoing: SON BORN JULY SEVEN BOTH WELL

Incoming: SERVICEMAN INFORMED BY FRIEND MOTHER DIED TWO MONTHS AGO STILL RECEIVING LETTER FROM HER REGULARLY INVESTIGATE

Outgoing: MOTHER DIED CANCER BREAST APRIL TWENTY SEVENTH SISTER FEARED SHOCK TO SERVICEMAN HAS BEEN WRITING IN MOTHERS NAME WILL WRITE IMMEDIATELY

Incoming: SERVICEMAN BEGS WIFE DISREGARD HIS LAST LETTER MAILED JULY SEVEN RECEIVED ONE HUNDRED FIFTEEN LETTERS FROM HER YESTERDAY

Outgoing: MESSAGE RECEIVED WIFE WILL WRITE

Incoming: SERVICEMAN REQUESTS HEALTH CONFINEMENT RE-
PORT WIFE

Outgoing: WIFE DIED CHILDBIRTH JULY TWENTY SEVEN SON
WELL WITH SERVICEMAN'S MOTHER-IN-LAW GET-
TING GOOD CARE MOTHER-IN-LAW WILL KEEP CHILD
UNTIL SERVICEMAN'S RETURN

The Home Service Corps workers made two contributions to the development of service delivery. First, extended help was offered to individuals and their
families. The intervention was problem focused but not timebound. The Home
Service Corps volunteer helped identify the problem and stayed with each family until it was resolved. Second, the volunteer did not just solve problems, but
also became a broker of services. He or she would often coordinate communications and requests for services between the family and the agencies that could
provide the help and support. The work often involved helping families communicate with the military (Dulles, 1950; Hurd, 1959).

After World War II, it became increasingly evident that many people needed
assistance to improve their quality of life. In the early 1960s, the federal government became increasingly involved in helping people in need.

The Impact of Federal Legislation: The Past and the Present

Several pieces of federal legislation were passed between the mid-1960s and the
late 1980s and modified in recent years. These legislative efforts recognize the
need for a case management process to provide social services to people in need.
Serving older adults, children with disabilities, and the families and children
who live in poverty requires the services that case management represents.

ELDERCARE

The Older Americans Act of 1965 (Public Law 89-73) focused on providing services for older individuals in order to improve their quality of life. Among its contributions to the development of case management was an emphasis on the
multiplicity of human needs. A summary of the act follows.

> To provide assistance in the development of new or improved programs
> to help older persons through grants to the States for community plan-
> ning and services and for training, through research, development, or
> training project grants and to establish within the Department of Health,
> Education, and Welfare an operating agency to be designated as the "Ad-
> ministration on Aging." (Public Law 89-73, 1965)

This act advanced case management by recognizing the need to coordinate
care. Section 101 of the act describes its goals and the services to be provided.
(See Box 2.2.) The services are designed to meet a variety of needs: financial,
medical, emotional, housing, vocational, cultural, and recreational.

BOX 2.2   The Older Americans Act of 1965
• • • • • • • • • • • •

Title I—Declaration of Objectives: Definition

Sec. 101. The Congress hereby finds and declares that, in keeping with the traditional American concept of the inherent dignity of the individual in our democratic society, the older people of our Nation are entitled to, and it is the joint and several duty and responsibility of the governments of the United States and of the several States and their political subdivisions to assist our older people to secure equal opportunity to the full and free enjoyment of the following objectives:

(1) An adequate income in retirement in accordance with the American standard of living.
(2) The best possible physical and mental health which science can make available and without regard to economic status.
(3) Suitable housing, independently selected, designed, and located with reference to special needs and available at costs which older citizens can afford.
(4) Full restorative services for those who require institutional care.
(5) Opportunity for employment with no discriminatory personnel practices because of age.
(6) Retirement in health, honor, dignity—after years of contribution to the economy.
(7) Pursuit of meaningful activity within the widest range of civic, cultural, and recreational opportunities.
(8) Efficient community services which provide social assistance in a coordinated manner and which are readily available when needed.
(9) Immediate benefit from proven research knowledge which can sustain and improve health and happiness.
(10) Freedom, independence, and the free exercise of individual initiative in planning and managing their own lives.

SOURCE: Older Americans Act of 1965, Public Law 89-73.

Today the need for support for care of the elderly has increased as the number of individuals in the United States over the age of 65 is projected to more than double by 2050 (Clubok, 2001). Plagued by chronic illness, depression, insufficient financial resources, and many other difficulties, this population is already placing demands on many families, communities, and the human service delivery system. Even though federal programs such as Social Security, Medicaid, and Medicare support many individuals, these resources need to be used effectively and efficiently. (See Box 2.3.) Case management will continue to be used in eldercare to serve these individuals with complex short- or long-term needs.

Client involvement, a basic principle of case management today, was also an important theme in the Rehabilitation Act of 1973 and its subsequent amendments through 1986. These pieces of legislation promoted consumer involvement

SSDI Program

Title II of the Social Security Act establishes the Social Security Disability Insurance Program (SSDI). SSDI is a program of federal disability insurance benefits for workers who have contributed to the Social Security trust funds and become disabled or blind before retirement age. Spouses with disabilities and dependent children of fully insured workers (often referred to as the primary beneficiary) also are eligible for disability benefits on the retirement, disability, or death of the primary beneficiary. Section 202(d) of the Social Security Act also establishes the adult disabled child program, which authorizes disability insurance payments to surviving children of retired or deceased workers or workers with disabilities who were eligible to receive Social Security benefits, if the child has a permanent disability originating before age 22.

Hereinafter in this policy brief, the term *SSDI* refers to all benefit payments made to individuals on the basis of disability under Title II of the Social Security Act.

SSDI provides monthly cash benefits paid directly to eligible persons with disabilities and their eligible dependents throughout the period of eligibility.

SSI Program

Title XVI of the Social Security Act establishes the Supplemental Security Income (SSI) Program. The SSI program is a means-tested program providing monthly cash income to low-income persons with limited resources on the basis of age and on the basis of blindness and disability for children and adults. The SSI program is funded out of the general revenues of the Treasury.

Eligibility for SSI is determined by certain federally established income and resource standards. Individuals are eligible for SSI if their "countable" income falls below the federal benefit rate ($512 for an individual and $769 for couples in 2000). States may supplement the federal benefit rate. Not all income is counted for SSI purposes. Excluded from income are the first $20 of any monthly income (i.e., either unearned, such as Social Security and other pension benefits, or earned) and the first $65 of monthly earned income plus one-half of the remaining earnings. The federal limit on resources is $2,000 for an individual and $3,000 for couples. Certain resources are not counted, including, for example, an individual's home and the first $4,500 of the current market value of an automobile.

The Ticket to Work and Work Incentives Improvement Act of 1999

On December 17, 1999, President Clinton signed into law the Ticket to Work and Work Incentives Improvement Act of 1999 (Public Law 106-170). Hereinafter in this policy brief, Public Law 106-170 will be referred to as *the Act*. The Act has four purposes [Section 2(b) of the Act]:

(continues)

(continued)

- To provide health care and employment preparation and placement services to individuals with disabilities that will enable those individuals to reduce their dependency on cash benefit programs.
- To encourage states to adopt the option of allowing individuals with disabilities to purchase Medicaid coverage that is necessary to enable such individuals to maintain employment.
- To provide individuals with disabilities the option of maintaining Medicare coverage while working.
- To establish a return-to-work ticket program that will allow individuals with disabilities to seek the services necessary to obtain and retain employment and reduce their dependency on cash benefit programs.

This Act improves work incentives under the SSDI and the SSI and expands health care services under Medicare and Medicaid programs for persons with disabilities who are working or who want to work but fear losing their health care.

SOURCE: Center on State Systems and Employment (RRTC). (2000). Policy Brief: Improvements to the SSDI and SSI Work Incentives and Expanded Availability of Health Care Services to Workers with Disabilities under the Ticket to Work and Work Incentives Improvement Act of 1999. Policy Brief, Vol. 2, No. 1. [On-line]. Available: <http://www.childrenshospital.org/ici/publications/text/pb2text.html>

Medicare

Medicare is a health insurance program for:

- People 65 years of age and older.
- Some people with disabilities, under 65 years of age.
- People with end-stage renal disease (permanent kidney failure requiring dialysis or a transplant).

Medicare provides hospital insurance. Most people do not have to pay for this insurance. Medicare also provides medical insurance and most people pay monthly for this insurance. There are several ways that individuals can access their Medicare benefits.

- The Original Medicare Plan—This plan is available everywhere in the United States. It is the way most people get their hospitalization and medical insurance benefits. They can go to any doctor, specialist, or hospital that accepts Medicare. Medicare pays its share and the recipient pays a share. Some services are not covered, such as prescription drugs.
- Medicare Managed Care Plans—These are health care choices (like HMOs) in some areas of the country. In most plans, individuals can only go to doctors, specialists, or hospitals that are part of the plan. Plans must cover all Medicare hospitalization and medical insurance benefits. Some plans cover extras, like prescription drugs. Individual out-of-pocket costs may be lower than in the Original Medicare Plan.
- Private Fee-for-Service Plans—This is a new Medicare health care choice in some areas of the country. Individuals may go to any doctor, specialist, or hos-

pital. Plans must cover all hospitalization and medical insurance benefits. Some plans cover extras like extra days in the hospital. The plan, not Medicare, decides how much you pay.

SOURCE: Medicare: The Official U.S. Government Site for Medicare Information. (2000). [On-line]. Available: <http://www.medicare,gov/Basics/WhatIs.asp>

Medicaid

Title XIX of the Social Security Act is a program that provides medical assistance for certain individuals and families with low incomes and resources. The program, known as Medicaid, became law in 1965 as a jointly funded cooperative venture between the Federal and state governments to assist states in the provision of adequate medical care to eligible needy persons. Medicaid is the largest program providing medical and health-related services to America's poorest people. Within broad national guidelines that the federal government provides, each of the states:

- Establishes its own eligibility standards
- Determines the type, amount, duration, and scope of services
- Sets the rate of payment for services
- Administers its own program

Thus, the Medicaid program varies considerably from state to state, as well as within each state over time.

SOURCE: Health Care Financing Administration. (2000). [On-line]. Available: <http://www.hcfa.gov/medicaid/mover.htm>

while serving individuals with severe disabilities (Rubin & Roessler, 2001), particularly in eligibility determination and plan development. Client satisfaction and the adequacy of services also received attention. Client involvement was further strengthened when the Rehabilitation Act Amendments of 1992 and 1998 were passed, emphasizing consumer choice and control in setting goals and objectives.

Public Law 94-142, Children with Disabilities, Education for all Handicapped Children Act of 1975, included an explicit case management process so as to treat the client as a customer. The client was to be involved in identifying the problem, given complete information about the results of the assessment of needs, and empowered to help determine the type of services delivered. The client also participated in the evaluation of the helping process and in any decision to terminate or redirect the activities (Jackson, Finkler, & Robinson, 1992).

The passage and subsequent implementation of this act serves as an excellent example of how federal legislation applies the case management process (Weil & Karls, 1985a). This public law, now known as the Individuals with Disabilities

Education Act (IDEA), guarantees appropriate public education to all children who have educational, emotional, developmental, or physical disabilities. Congress found that the educational needs of more than 8 million handicapped children in the United States were not being met. According to the statute, state and local educational agencies would provide services to these children, and the federal government would assist with the funding. All handicapped children are to have access to free public education, including special education and other services to address their unique educational needs.

One tool to assist with planning, implementation, and evaluation was the *individualized educational program* (IEP). The IEP articulates a plan of intervention for each child based on goals and recommended intervention strategies. Although educational agencies have implemented the IEP in numerous ways, it has always been critical that the case manager assume the leadership role of the team of participants who develop the IEP.

On June 24, 1997, President Bill Clinton signed into law the IDEA amendments. These amendments reflect changes in the values and process of case management and the planning and implementation of the IEP, focusing on four important areas: strengthening parental participation, accountability for student participation in general education, remediation and behavior problems addressed within the educational environment, and preparation of students for independent living.

Two of these issues relate directly to the case management process. First, as the parental role changed, the emphasis on client empowerment was strengthened. Parents representing the interests of the student now have first-line signature on the IEP. Parents must be explicitly consulted about each component of the plan and must signify their agreement in writing. Parents must be present at each IEP meeting and must approve every change to the IEP.

The second amendment that enhances the case management process reinforces the emphasis on client strengths. For the first time, there is a section on the IEP that asks for children's strengths to be listed, and it is the first assessment information on the IEP. These strengths are used to plan the student's program, which will be implemented throughout the year. They will also guide the transition plan that focuses on educating the students for independent living.

FAMILIES AND YOUNG CHILDREN

In 1988, the Family Support Act was passed. As described here, the act mandated that case management be applied to the process of serving those who were deemed eligible. This marked new status for the case manager. The act was passed with the express goal of increasing the economic self-sufficiency of families who receive Aid to Families with Dependent Children (AFDC). It increased the level of child support enforcement and added a new welfare-to-work program. A new Job Opportunity and Basic Skills program (JOBS) replaced the previously funded Work Incentive Program (Hagen, 1994). One notable characteristic of the act is the explicit authorization of the case management function in implementation. Education, training, employment-related services, and case management were to be available to JOBS clients and their children. Specifically, the purpose of JOBS was to make sure that needy families with children receive the education, training, and employment to help them avoid long-term welfare. The responsibility for implementing JOBS resides with the state welfare agency.

The Family Support Act of 1988 endorses the importance of the case management function in coordinating services for clients—in this case, young mothers and their children. In the legislation, case management is defined as a process that "must be responsible for assisting the family to obtain any services which may be needed to assist effective participation in the program." In addition, the act encouraged the development of new models and definitions of case management, since states could design their own case management systems (Hagen, 1994).

In 1996, AFDC was replaced by the Personal Responsibility and Work Opportunity Act. This new welfare legislation required that young mothers receive financial support for two years while they get vocational education and training to join the work force. Case managers became a key component in these welfare-to-work programs as they helped develop and coordinate the plan that moves young mothers to self-sufficiency. For example, New Jersey counties Atlantic and Cape May developed a collaborative case management model that included the New Jersey Department of Labor, Department of Human Services, and the Department of Health and Senior Services. The collaboration included representatives from state and local government agencies as well as more than 42 community-based organizations. They developed "One Ease E Link," an electronic linkage that serves as the basis for the elaborate collaborative case management and referral system (Welfare & Workforce Development Partnerships, 2000).

Just as social legislation was a major factor in the development of case management in the 1960s, 1970s, and 1980s, the advent of managed care during the 1980s has expanded the range of case management in the 1990s. The next section describes managed care and explores its impact on service delivery.

 # The Impact of Managed Care

The emergence of managed care as a model of health care delivery has increased the demand for case management services and provided new models and definitions of service delivery. To understand its impact, one must first grasp what managed care is.

History of Managed Care

Until the 1930s, most medical care in this country was provided on a fee-for-service basis. This means that a patient would be assessed a fee for each health or mental health service provided by a professional. For example, when Mrs. Fowler goes for her annual checkup, she receives a bill for the doctor's consultation time, the tetanus injection, and the EKG. In the early 1930s, physicians implemented prepaid group plans or managed plans for medical services. This was an alternative way of organizing medical care. The basic concept of a prepaid plan was to guarantee a defined set of services for a negotiated fee. On such a plan, Mrs.

Fowler would pay a yearly fee that covered a set of services such as those provided at her annual checkup.

The growth in prepaid group plans was relatively slow until the 1970s. Then the Health Maintenance Organization Act of 1973 (Public Law 93-222) allowed managed medical plans to increase in number and expand the numbers of patients being served (MacLeod, 1993). The prevalence of managed care is now commonly regarded as being connected to the rising cost and decreasing quality of health care and mental health care. The escalation of costs reflects many trends: improved technology, shifting of costs from nonpaying patients to paying patients, an older population, high expectations for a long and healthy life, increased administrative costs, and varying standards of efficiency and quality care (Kongstvedt, 1993).

Defining Managed Care

There are several ways to define **managed care**. First, the term may simply refer to an organizational structure that uses prepayment rather than fee-for-service payment. Second, it can designate the array of different payment plans, such as prepayment and negotiated discounts. It may also imply the inclusion of quality assurance practices, such as agreements for prior authorization and audits of performance (Dziegielewski, 1998). Third, *managed care* may be used to refer to the policy of restricting clients' access to providers such as physicians and other health professionals. Instead, the providers or professionals are paid a flat fee to provide service to a certain group of patients or clients. Most simply stated, managed care is an agreement that health providers will guarantee services to clients within specified limits. The restrictions are intended to improve efficiency of services (Dziegielewski, 1998). According to the same authors, the goals of managed care are as follows:

- To encourage decision makers (providers, consumers, and payers) to evaluate efficiency and priority of various services, procedures, and treatment
- To use the concept of limited resources in making decisions about services, procedures, and treatment
- To focus on the value received from the resources as well as the lower cost

Models of Managed Care

Three types of managed care models have evolved to meet the goals stated in the previous paragraph: health maintenance organizations (HMOs), preferred provider organizations (PPOs), and point of service (POS). Each has a particular strategy for maintaining cost and ensuring quality.

HEALTH MAINTENANCE ORGANIZATIONS

The HMO managed care model, which is the most structured and controlled, emphasizes positive health promotion. **HMO** is a generic term covering a wide range of organizational structures; it is distinguished from traditional fee-for-service health care systems by combining delivery and financing into one system. According to federal law, an HMO must have three characteristics (Joint Interim Committee on Managed Care, 2000):

- "An organized system for providing healthcare or otherwise assuring healthcare delivery in a geographic area.
- An agreed upon set of basic and supplemental health maintenance and treatment services.
- A voluntary enrolled group of people. The HMO assumes the financial risk for providing the contracted services."

There are advantages and disadvantages to the HMO model. One immediate benefit is that both the services available and the cost of providing them are constantly monitored by the HMO. Physicians and other health professionals must establish a rationale for services, procedures, and treatment recommended, and their rate is monitored. Client spending is also monitored, since enrolled members can receive services only from the professionals participating in the plan. In some instances, the clients must obtain preauthorization from someone outside the plan. Through its leverage at the site of services, the HMO can control utilization and improve the efficiency of service delivery (Comarow, 1998).

From the client's perspective, the site-of-services restrictions also represent the greatest disadvantage of an HMO. Clients do not like limits on their use of providers; they wish for more freedom to choose. In response to members' demands for freedom to choose their own providers, two other managed care systems have emerged: PPOs and POS (Boland, 1993).

PREFERRED PROVIDER ORGANIZATIONS

The term **preferred provider organization (PPO)** does not describe any single type of managed care arrangement. Rather, this plan falls between the traditional HMO and the standard indemnity health insurance plan. The following characteristics apply to PPOs (Joint Interim Committee on Managed Care, 2000):

- Contracts are established with providers of medical care.
- These providers are referred to as preferred providers.
- The benefit contract provides significantly better benefits for services received from preferred providers.
- Covered persons are allowed benefits for non-participating providers' services.

PPOs point with pride to their prompt payment of claims. The providers accept a negotiated discount, which represents the PPO fee, and they do not bill patients an additional amount. Based on the negotiated fee, both the clients and the PPO can anticipate their costs, and providers can anticipate their income. From the providers' perspective, they are assuming a business risk in terms of the fees that they agree to accept. On the other hand, they expect to increase the number of patients under their care. Many providers also maintain independent medical practices.

POINT OF SERVICE

The third option of managed care offered today is the POS. It is often adopted by traditional HMO members who want more flexibility than the HMO or the PPO provide. The **point of service** plan is characterized by the following features:

- Customers are allowed to use out-of-plan providers, and they receive reduced coverage.
- To participate in the POS plan, clients pay higher premiums, higher deductibles, and a higher percentage of the medical fees (Curtiss, 1990).
- In this system, clients are encouraged to use the providers in the managed care system, but they do not use all their benefits if they choose medical care outside the system.

Managed care has emerged as a response to the fact that employers, governments, payers, clients, and providers are all seeking ways of containing health care costs. All three plans presented emphasize management of medical cases, review and control of utilization, and incentives for or restrictions on providers and clients to reduce costs and maintain quality. It is still too early to judge whether managed care systems are achieving these goals, but some clear advantages and disadvantages have emerged. Proponents of the system point to the following advantages of managed care.

- Providers, clients, and payers involved in health care are beginning to prioritize among services, procedures, and treatments provided.
- Providers are more thoughtful about any plan of action prescribed for the client, since they must provide justification for each component of the plan.
- Efficiency of service delivery improves because services are to be provided in the shortest time needed to meet the goals established.
- From the client's perspective, there is a single, coordinated point of entry into the system.
- Resources are saved.
- Resources are spent according to priorities.

Those who question the virtue of the managed care system focus their concerns in two areas: the quality of services delivered and the efficiency of service. They see the following disadvantages of delivering services through managed care.

- Professionals do not believe that they are offering the best services available, because they are constrained by resource limitations.
- Managed care staff are making judgments about the suitability of proposed treatment without adequate training or professional knowledge.

- Services are not delivered in a timely manner because of the extra layer of bureaucracy.
- Managed care organizations require paperwork that limits the amount of time the professional can give the client.
- Clients worry about the quality of the care they are receiving.
- Clients do not have access to all the services they believe they should have.
- Clients cannot choose their service providers.

In response to the professional and patient or client frustrations with managed care, the concept of a bill of patient rights has evolved. These rights would include the following: the right of appeal when denied services, access to specialists when needed, use of doctors outside the health plan, receiving understandable explanations of coverage regulations (Fromkin, 1999). In addition, standards for evaluating managed care organizations have been developed in an effort to define standards of quality. The National Committee for Quality Assurance (NCQA), an organization that accredits managed care plans, cite the following characteristics of quality indicators: prevention, access to care, member satisfaction, and physician credentials (Comarow, 1998).

In spite of the criticisms raised, managed care is no longer just one alternative in the health care delivery system; it is part of the structure of service delivery. Managed care is also being adopted by the human service delivery system. This shift to managed care first influenced human services that were directly tied to health care, such as substance abuse and care for the elderly. In these areas, the social service or mental health service delivered has been restricted by the same provider agreements as described in the HMO, PPO, and POS plans. The use of managed care in human services will continue to influence the delivery of services. Throughout this text you will see how managed care has influenced the practice of case management. One example of the use of managed care is described by the Child Welfare League, an association of more than 1,000 public and private not-for-profit agencies that assist 3.5 million children and their families each year. According to the Child Welfare League of America (2000),

> there are many definitions of managed care. Generally, the term refers to the use of a variety of approaches intended to balance the cost of services with quality and access. Because resources are limited, managed care is designed to reconcile the provision of care to each individual with the resources available to serve the entire pool. Cost effectiveness is achieved by efficiently delivering the most appropriate services to each individual served.
>
> Managed care systems balance cost, quality, and access by utilizing the following tools: pre-authorization for care; gate keeping and utilization review; use of standardized practice guidelines; increased utilization of information technology to manage data effectively; built in financial risks and incentives for providers; and outcome-based contracting.
>
> If designed and priced correctly, a managed care system has the potential to improve quality and accountability; create a system that is

data-driven; reduce service gaps and duplication; and ensure that clients get the services, the right amount of services they need, when they need them. No more, no less!

Increasingly, public administrators are integrating managed care techniques and tools into the delivery of child welfare services. This process often involves sharing financial risk and case management responsibility between the public agency and the private agencies or managed care entities. These techniques can be applied to the full range of child welfare services. These include, but are not limited to: foster care, adoption, family preservation, residential treatment for children and adolescents, emergency shelter, parenting education and other preventative skill building, substance abuse and mental health treatment for children and their families, kinship care, and therapeutic foster care.

 ## Expanding the Responsibilities of Case Management

Given the various perspectives on case management, its historical development, and the impact of managed care, it will be necessary to revise the case management process for the future. Shifts are evident in client involvement, the roles of the helper, and the emphasis on cost containment. Historically, the case management process has emphasized coordination of services, interagency cooperation, and advocacy, but the process of service delivery is expanding. Other trends have emerged from federal legislation, including coordination of care, integration of services, and the client as a customer. More recently, professionals providing services have been encouraged to empower clients, contain costs, and ensure quality services. These shifts in emphasis are reflected in the roles and responsibilities of the people delivering service today.

 ## Chapter Summary

This chapter establishes a historical context for case management by describing four perspectives on case management that have evolved in the past 30 years. A brief history of case management, the impact of managed care, and expanding case management responsibilities complete the historical perspective.

The four perspectives on case management illustrate the recent development of the process. Case management as a process was recognized during deinstitutionalization when limited assessment was followed by placement. Client involvement, the second perspective, evolved during the 1980s as clients were encouraged to become partners in the process. A shifting emphasis from management to coordination characterized the third perspective, which described the role of the case manager. The final perspective concerns utilization review and cost–benefit analysis that resulted from spiraling costs and health care reform.

The history of case management traces the use of advocacy, data gathering, recordkeeping, and cooperation. Federal legislation, such as the Older Americans Act of 1965, the Individuals with Disabilities Education Act, and the Family Support Act, further developed case management.

By the 1970s, managed care was seen as the answer to the rising cost and decreasing quality of health care. It is now part of the structure of service delivery, and it affects both social and mental health services.

Finally, case management responsibilities continue to evolve. Empowering clients, containing costs, and ensuring quality services are reflected in case management responsibilities today.

Chapter Review

◆ *Key Terms* ....

Deinstitutionalization	*PPO*
Social diagnosis	*HMO*
Fee-for-service	*POS*
Managed care	

◆ *Reviewing the Chapter* ....

1. Trace the development of the case management function since the 1970s.
2. Describe the shift in the use of the term *case management* that occurred in the 1980s.
3. Why have *service coordination* and *care coordination* become common terms for case management?
4. How has the managed care movement influenced case management?
5. Describe the influence on case management of early organizations such as the Massachusetts School of Idiotic and Feebleminded Youth and the Board of Charities.
6. What emerging case management functions are illustrated by Hull House and the Henry Street Settlement House?
7. How did immigrant populations benefit from the form of case management practiced in the settlement houses?
8. Describe the case management functions of the American Red Cross.
9. What has been the impact of federal legislation on case management?
10. How did federal legislation contribute to the evolution of case management?
11. Trace the development of managed care in the 20th century.
12. Explain the various ways the term *managed care* is defined.
13. Identify and describe the three models of managed care.

◆ *Questions for Discussion* ....

1. What part of the history of case management most influences the process today? Cite the reasons for your answer.

2. Why do you think case management developed in the United States over the past century?
3. If you could choose a managed care organization with which to work, what criteria would you use to make your decision?
4. What do you predict will be the future of case management in the United States?

References

Addams, J. (1910). *Twenty years at Hull House*. New York: New American Library.
Boland, P. (1993). Evolving managed care organizations and product innovation. In P. Boland (Ed.), *Making managed care work: A practical guide to strategies and solution* (pp. 151–192). Gaithersburg, MD: Aspen Press.
Brieland, D. (1990). The Hull House tradition and the contemporary social workers: Was Jane Addams really a social worker? *Social Work, 35*(2), 134–138.
Child Welfare League of America. (2000). Managed Care Institute. [On-line]. Available: <http://www.cwla.org/programs/managedcare>
Clubok, M. (2001). The aging of America. In T. McClam & M. Woodside (Eds.), *Human service challenges in the 21st century* (pp. 352–358). Birmingham, AL: EBSCO.
Comarow, A. (1998). Behind HMO ranking. [On-line]. Available: <http://www.usnews.com/usnews/nycu/health/hehmobeh.htm>
Curtiss, F. (1990). Managed care: The second generation. *American Journal of Hospital Pharmacy, 47*(9), 2047–2052.
Dulles, F. (1950). *The American Red Cross*. New York: Harper.
Dunst, C., & Trivette, C. (1989). An enablement and empowerment perspective of case management. *Topics in Early Childhood Special Education, 8*(4), 87–102.
Dziegielewski, S. (1998). *The changing face of health care social work*. New York: Springer.
Fromkin, D. (1999). Backlash builds over managed care. [On-line]. Available: <http://washingtonpost.com/up-srv/health/policy/managedcare/overview/htm>
Hagen, J. (1994). JOBS and case management: Developments in 10 states. *Social Work, 39*(2), 197–205.
Hurd, C. (1959). *The compact history of the American Red Cross*. New York: Hawthorne.
Jackson, B., Finkler, D., & Robinson, C. (1992). A case management system for infants with chronic illnesses and developmental disabilities. *Children's Health Care, 21*(4), 224–232.
Joint Interim Committee on Managed Care (2000). Glossary of HMO terms. [On-line]. Available: <http://www.senate.state.mo.us/mancare/terms.htm>
Kongstvedt, P. (1993). *The managed health care handbook*. Gaithersburg, MD: Aspen Press.
Kuno, E., Rothbard, A., & Sands, R. (1999). Service components of case management which reduce inpatient care use for persons with mental illness. *Community Mental Health Journal, 35*(2), 153–167.
Lieberman, F. (1990). The immigrants and Mary Richmond. *Child and Adolescent Social Work, 7*(2), 81–84.
Lourie, N. (1978). Case management. In J. Talbott (Ed.), *The chronic mental patient* (pp. 159–164). Washington, DC: American Psychiatric Association.
MacLeod, G. (1993). An overview of managed healthcare. In P. Kongstvedt (Ed.), *The managed health care handbook* (pp. 1–3). Gaithersburg, MD: Aspen Press.

Montalvo, F. (1982). The third dimension in social casework: Mary E. Richmond's contribution to family treatment. *Clinical Social Work Journal, 10*(2), 103–112.

Ozarin, L. (1978). The pros and cons of case management. In J. Talbott (Ed.), *The chronic mental patient* (pp. 165–170). Washington, DC: American Psychiatric Association.

Pittman-Munke, P. (1985). Mary E. Richmond: The Philadelphia years. *Social Case Work, 66*(3), 160–166.

Polacheck, H. (1989). *I came a stranger: The story of a Hull-House girl.* Chicago: University of Illinois Press.

Residents of Hull House. (1985). *Hull House maps and papers.* New York: Thomas Y. Crowell.

Richmond, M. (1917). *Social diagnosis.* New York: Russell Sage.

Rubin, S. E., & Roessler, R. T. (2001). *Vocational rehabilitation process.* Austin, TX: PRO-ED.

Trattner, W. I. (1994). *From poor law to welfare state: A history of social welfare in America.* New York: Free Press.

Trustees of the Massachusetts School for the Feebleminded. (1919). *Seventy-second annual report.* Boston: Wright and Potter.

Trustees of the Massachusetts School for the Feebleminded. (1920). *Seventy-third annual report.* Boston: Wright and Potter.

Wald, L. (1915). *The house on Henry Street.* New York: Dover Press.

Weil, M., & Karls, J. (1985a). *Case management in human service practice.* San Francisco: Jossey-Bass.

Weil, M., & Karls, J. (1985b). Key components in providing efficient and effective services. In M. Weil & J. Karls (Eds.), *Case management in human service practice* (pp. 29–71). San Francisco: Jossey-Bass.

Welfare & Workforce Development Partnerships. (2000). Atlantic and Cape May Counties, New Jersey. [On-line]. Available: <http://wtw.doleta.gov/wwpartnerships>

Winsor, J. (Ed.). (1881). *The memorial history of Boston 1630–1880* (Vol. IV). Boston: Ticknor.

Woods, R., & Kennedy, A. (1911). *Handbook of settlements.* New York: Russell Sage.

Chapter Three ···

Models of Case Management

I *did a lot of my work the last four years in rehab. I had a lady who had a stroke. She did well on rehab but she needed support. She couldn't go home independently. The son was not in the picture, so we developed a network like meals on wheels. . . . You know, we tried to come up with therapy in the home and access to transportation . . . that sort of thing. Then we sent her home and I would follow up with phone calls.*
   —Judith Slater, Kennesaw State University, Kennesaw, GA, personal communication, October 14, 1995

I *work with an agency that serves adults with developmental disabilities. Our age group ranges from 16 to at least 78 as our oldest. We have a residential program. We also have a day program. We have independent living. We do everything. I am learning that. I am a service coordinator, which we used to call case managers.*
   —Kim Ehlers, Mitchell Area Adjustment Training Center, Parkston, SD, personal communication, October 14, 1995

R *elease plans are forms the inmates fill out about where they want to live and work once they get out. The parole officer investigates both of these. . . . We check out all of that to be sure there is room at the proposed residence and that there is not a conflict there. As far as the job, we want to make sure it is a legitimate business. . . . If we turn it down, we notify the institutional parole officer wherever the inmate is that the plan is rejected for whatever reason. We ask them to resubmit another plan . . . it is up to them to resubmit this.*
   —Angela LaRue, Office of Parole, Knoxville, TN, personal communication, October 18, 1995

This chapter introduces the roles that helpers assume within the context of case management. We also present models of case management, reflecting the creative ways in which services are delivered.

   The preceding quotations relate to both the roles that case managers perform in service delivery and the models used to deliver services. Judith Slater, working in hospital rehabilitation in Georgia, supports her clients as a coordinator, expediter, planner, and problem solver. Her agency provides services both in the hospital and after hospital discharge. The Mitchell Area Adjustment Training Center in South Dakota provides services in a residential program, a day program, and independent living. The program makes a long-term commitment to clients, regardless of their abilities or status. This requires flexibility of roles to accommodate the needs of the client. Working from a very different model, Angela LaRue's roles in the parole office are defined by the state and shaped by her large caseload. Since the parolees assume most of the responsibility for themselves, Angela is primarily a recordkeeper, monitor, and problem solver.

For each section of the chapter, you should be able to accomplish the following objectives.

ROLES IN CASE MANAGEMENT
• Identify the roles in the case management process.

MODELS OF CASE MANAGEMENT
• List reasons that it is important to understand the different models of case management.
• Name the three models of case management.
• Illustrate each model with an example.
• Compare and contrast case management and problem solving.

 Roles in Case Management

Each model of case management requires a number of roles to meet the goals established. Among the roles are linking clients with services, providing services, and serving as an advocate for the client. The following roles are integral to the provision of case management services.

Advocate

An **advocate** speaks on behalf of clients when they are unable to do so, or when they speak but no one listens. The case management process presents many opportunities for advocacy. Working at various levels, the case manager represents the interests of the client, helping to gain access to services or improve their quality. At the organizational level, the case manager influences the policies that control eligibility and access to services. At the legislative level, case managers can work to influence government policies and programs that serve the needs of their clients, which include addressing issues of inequality and discrimination. Case managers also help clients to become advocates for themselves and their families. This is one way to empower clients.

Jim was an advocate for Bryan, a 19-year-old recently admitted to an inpatient program for treating substance abuse. Bryan had participated twice before in a short-term inpatient program, and each time he had returned after six months. As Jim, his care coordinator, evaluated Bryan's past history, it became obvious that the previous treatment had not been effective. Jim petitioned the managed care team and the alcohol and drug treatment team to develop an individualized program for Bryan. Although Jim's petitions are not always successful, he hopes the managed care team will be persuaded by his arguments.

Broker

As a **broker**, the case manager links the client with needed services. Once the client's needs are clear, the broker helps the client choose the most appropriate service and negotiates the terms of service delivery. In this brokering role, the case manager is concerned with the quality of the service available and any difficulties the client may have in accessing it.

When Jo Sinclair assumed the brokering responsibilities for her elderly mother, the job was not an easy one. Jo had returned home from college at the end of her sophomore year to find her mother in poor mental health. Jo's father had left home the previous month, and her mom had sunk into deep depression. As the broker, Jo arranged appointments for her mother to see a psychologist, a physician, and a lawyer. When Jo needs professional help, she calls the Office on Aging; members of their client services staff are available for consultation. The broker role is a difficult one for Jo since she is unaware of availability of services and the terms of eligibility.

Coordinator

Many clients have multiple problems and need more than one service to meet their needs. In the role of **coordinator**, the case manager works with other professionals and agency staff to ensure that services are integrated. The case manager must know the current status of the client, the service delivered, and the progress being made. Monitoring the client's progress and interfacing with professionals is an important role for the case manager.

Jamie Wolfenbarger assumes the coordinator role for the local hospital's long-term care clients. In this role, she plans the aftercare for patients who will require long-term care. She coordinates previously unrelated services performed by professionals from different agencies. For Rose Woodson, a patient soon to be released from the Cardiac Observation Unit, Ms. Wolfenbarger has arranged home health care visits once a day, a housekeeper to clean twice a week, meals to be delivered at noon each day, and special ambulatory equipment. Ms. Wolfenbarger will contact these professionals each week for the next month for feedback on Ms. Woodson's progress.

Collaborator

Helping professionals must establish and maintain good working relationships with other service providers. As a **collaborator**, the case manager knows the human service community and has established a network of other professionals. Case managers also collaborate with other professionals during team staffings and program planning.

LaTonya Welch works for a disaster recovery agency. She has colleagues and collaborators all over the United States, and she has begun to gather professional contacts worldwide. Her primary responsibility is to provide services in her jurisdiction to individuals, families, neighborhoods, and regions that have experienced a crisis—especially one brought about by a natural disaster. Her secondary responsibility is to provide help in crisis situations elsewhere in the nation. Last month she even worked on an international relief effort that collected clothing and air-freighted it to Chile after a volcano erupted there. Ms. Welch has to have a network of colleagues who can help put together services at a moment's notice.

Community Organizer

A case manager, acting as a **community organizer**, helps agencies work together to assess the needs of the community and plan how the local human service delivery system will meet those needs. It is sometimes difficult for organizations to cooperate in an atmosphere characterized by competition and limited resources. However, many forward-looking organizations have come to serve their clients more effectively once they have established good working relationships with other agencies in the community.

In his role as chief of staff at the local hospital, Dr. Chin Lee is a community organizer. He is particularly interested in meeting the needs of mentally ill adults and teens. At present he chairs a task force to assess the size and characteristics of the population of young people and adults with mental illness in the community.

Consultant

Often an outside professional can help solve case management problems. An organization may need assistance with such matters as cost analysis, quality control, and organizational structure. A **consultant** may have the expertise to identify the problem, study it, and make recommendations. Consultants can also assist with the case management of individual clients when special information or expertise is needed. This is especially true in small agencies that employ only generalist case managers.

Ann Marsella is a well-known expert on the treatment of young children with developmental disabilities. She is often called in for consultation on particularly challenging cases. Her expertise is in the legal and ethical aspects of serving these children and their families; she is respected for her ability to clarify a situation's ethical issues and the logical consequences of the proposed alternatives.

Counselor

The case manager who is a **counselor** or therapist maintains a primary relationship with the client and his or her family. Having a thorough understanding of the client's mental health and medical history, this professional can tell what aspects of his or her current situation support or discourage progress.

David Tanaka maintains therapeutic relationships with 15 clients for whom he also serves as a caseworker. He sees each client once a week for an hour and talks weekly with other professionals involved in each case. He expects to retain these 15 clients for the next two or three years, without increasing his caseload.

Evaluator

Evaluation is performed to determine the client's functioning and to assess service provision. In medical, psychological, financial, social, and vocational areas, the **evaluator** collects information from the client and from other professionals. The information is compiled and recommendations for treatment made.

Meena Shah is a case manager who works with a team of professionals, including a home health nurse, a psychiatrist, and a pharmacist, to conduct in-home assessments of elderly clients in the community. She is trained to complete a comprehensive assessment of functioning and to develop a treatment plan.

Evaluation also occurs during the implementation of the treatment plan. The case manager must evaluate the effectiveness of the plan in meeting the client's goals. This can result in minor or major changes to the plan when appropriate. Evaluation is also important when assessing the effectiveness of services and determining if outcomes have been met. It includes relating the plan to the case notes and the documentation of outcomes.

Renata Vennearo works in a school setting; she is an expert in working with children who have traumatic brain injury. One of her primary responsibilities is to manage the child's transition from hospital to school—a difficult matter because the condition is not neurologically stable. Ms. Vennearo constantly monitors the plan and reassesses needs as the client's needs change. Educational programming for these children must be more flexible than for other special education students.

Expediter

Clients encounter many difficulties in the human service delivery system. As **expediter**, the case manager helps the client get through problems of duplicated services, ineligibility, seemingly closed doors, poor service quality, and irrelevant services. The expediter ensures that services are delivered efficiently and effectively. As an expediter, the case manager also performs the advocacy function described earlier.

Judy Yow just called her client, Ms. Vanderhoff, following up after a physical therapy session. Ms. Vanderhoff goes to therapy once a week to maintain flexibility and to promote muscular development in her arms and legs. For the second Monday in a row, Ms. Vanderhoff had reported to the therapy center only to find it closed. Acting as expediter, Ms. Yow calls the therapy center and learns that the physical therapist had been in an automobile accident, and the center was unable to find a replacement. Ms. Yow spends the next hour locating another physical therapist for Ms. Vanderhoff on a short-term basis.

Planner

One of the primary responsibilities of the case manager as **planner** is preparing for the service or treatment that the client is to receive. Planning is directly connected to the findings of the assessment phase of case management. This includes setting goals, determining outcomes, and implementing the plan with input from the client, family members, other professionals, and other agencies. The case manager's planning role begins in the early stages of the helping process and continues until services are terminated. Planning may include a transition period until the client is able to manage his or her own case.

Tony Nix is in her first year as a case coordinator for the state parole board. She works with juvenile offenders after they have been paroled. Her first interaction with her clients occurs before they are released, and she develops a plan for their integration into their home environment. One of her greatest challenges is to plan for the first weeks after release since these young people often believe that being released means they can do anything they wish. To solidify their commitment to the stated goals and outcomes, she includes them in the planning process.

Problem Solver

The goal of the **problem solver** is to make clients self-sufficient by helping them determine their strengths, find alternatives to their current situations, and learn to solve their own problems. One area of problem solving is clarifying the roles of the client, the family, the caregiver, and the case manager. Disagreements about services, the direction of case management, or the plan often lead to conflicts. The case manager is continually involved in problem solving; many problems arise unexpectedly, and time must be allotted each day for them.

Sonja McCreless has always admired her direct supervisor, Jim Fitzpatrick, because he is an expert problem solver. She remembers his work with a very difficult client, Sue D'Ambrosio, who was scattered and unfocused. Jim Fitzpatrick was able to work with Sue to determine her own strengths, which included accomplishing very short-term tasks. So he modified the case management process into small tasks while clearly spelling out both of their responsibilities. Together they decided what the outcomes were to be and what behavior was acceptable, guided by the restrictions of the program. Sue responded positively to structured problem solving and eventually learned how to use the process without the guidance of her case manager.

Recordkeeper

Throughout service delivery, it is necessary to document assessment, planning, service provision, and evaluation. As a **recordkeeper**, the case manager maintains detailed information relating to all contracts and services. This is important for providing long-term care, communicating with other professionals and agencies, and monitoring and billing for services. Good documentation constitutes the linking element in the case management process. Many electronic systems of information management can help record, track, plan, monitor, and evaluate client progress, but the key is the quality of the data entered. This type of recordkeeping is essential for program evaluation. Agencies and individual case managers can determine if goals are being reached and if quality services are being provided. These data help to suggest if changes to organization, staffing, or service delivery need to be made.

Eli Brawley works with families and children having severe medical problems. He makes detailed computer records of his activities. Each client has a file that contains a record of every interaction and action taken for or with that client. This serves as the official record of service delivery, as well as the basis for accountability and quality assurance evaluations.

System Modifier

In human service systems, case managers often give advice or alter systems so that agencies can work better together. They also evaluate how well the collaboration among agencies is helping the target populations. **System modifier** responsibilities are usually assumed by case managers who have administrative responsibilities. They possess the authority to change agency policy or redirect priorities, in addition to having the respect of the community and their colleagues.

Heather Lagenly is the director of the Office on Aging. She also serves on the board of the Professional Society for Geriatric Care, a group of community professionals committed to providing quality services for the elderly. One goal of the society is to make sure that the system offers a variety of services to the target population, does not duplicate services unnecessarily, and provides services in a coordinated way. In its sixth year, the group is deciding how to deal with the growing number of elderly people who wish to be cared for in their own homes.

 ## Models of Case Management

This section introduces three modern models of case management. For several reasons, it is important to understand the different models. First, the existence of three separate models demonstrates that service delivery can occur in a variety of ways; it is a flexible process. Second, the different goals that characterize each model provide a perspective on the case manager's responsibilities, roles, and length of involvement with the client. Third, each model has particular strengths and weaknesses. Knowing these models helps determine in which particular settings each model is most relevant.

The three models presented have their bases in roles, organizations, and responsibilities (see Table 3.1). The **role-based case management** model centers on the roles the case manager is expected to perform. One case manager may act primarily as a broker of services, whereas another is a therapist who has occasional brokering and coordinating responsibilities. Another case manager may concentrate on cost containment and cost efficiency. Judith Slater's chapter-opening quote exemplifies the broker model; her major responsibility is linking the client to needed services.

Organization-based case management is a model that focuses on providing a comprehensive set of services and meeting the needs of clients with multiple problems. This model can be applied to work in a variety of situations, such as a comprehensive service center, a psychosocial rehabilitative center, or an interdisciplinary team approach. A quotation at the beginning of the chapter described an agency that serves individuals with developmental disabilities; the organization-based model applies in this case. The agency provides a range of services, all available in one location. Some clients live at the center.

In the **responsibility-based case management** model, the case management function may be performed by family members, a supportive care network, volunteers, or the client. In the third chapter-opening quotation, Angela LaRue described how the parolee assumes responsibility for a release plan. At her agency, the client has many responsibilities and is empowered to find a job and housing.

TABLE 3.1 MODELS OF SERVICE DELIVERY

	Models		
	Role-based	*Organization-based*	*Responsibility-based*
Examples	Generalist	Comprehensive service center	Family
	Broker	Interdisciplinary team	Supportive care
	Primary therapist	Psychosocial rehabilitative center	Volunteer
	Cost containment		Client as case manager

In the material that follows, each model is described according to the following characteristics: the goals of the process, the responsibilities and roles of the case manager, the length of the process, and the strengths and weaknesses of the model. Illustrations of each model follow the description of its characteristics.

Role-Based Case Management

The role assumed by the designated case manager characterizes this model of case management. Roles may vary according to the function and the services provided (Austin, 1981; Jackson, Finkler, & Robinson, 1992; Mullahy, 1998; Weil & Karls, 1985). Following the characteristics of this model, we present four illustrations: the case manager as a generalist, a broker, a therapist, and a cost container.

Goal   The case manager attempts to meet all the needs of the client through a single point of access. This may require the case manager to serve as the link to a variety of needed services, to be the provider of therapeutic care, or to monitor the efficiency and quality of services.

Responsibilities   The case manager assumes a broad set of responsibilities, including intake interviewing, data gathering, planning, linking to services, coordinating or delivering services (or both), referral, and evaluation.

Primary Roles   The roles can include broker, collaborator, coordinator, counselor, evaluator, expediter, planner, problem solver, and recordkeeper.

Length of Involvement   The duration of case management varies according to the complexity of the client's needs, the limitations managed care places on service delivery, and the need for short-term or long-term care.

Strengths   In many cases, there is a single point of access for the client. The case manager assumes the various roles as needed. Together they identify problems and develop a plan of services. The case manager remains closely involved as both provider and coordinator of services. This involvement promotes a strong relationship with the client, who has regular access to services. The case manager also provides assistance with financial aspects of the medical and mental health systems.

Weaknesses   Some agencies limit the services they provide; case managers may have large caseloads, no backup, and limited time for community involvement. Limited services may also mean more referrals, incomplete assessment, and a narrowing of the focus to only one perspective on service delivery. Where the case manager's role is oriented toward cost containment, there may be a focus on managed care concerns or a standard of care that does not correspond to the client's needs.

ILLUSTRATIONS

The first illustration of role-based case management involves the case manager as a *generalist*. This role, widely used in human service delivery, focuses primarily on providing the services that can be delivered by a helper with knowledge and skills applicable to a range of clients in various settings. Brokering or linking the client to other services occurs infrequently.

> Azzurra Given's life revolved around Florida State University, where she taught French and Italian literature for more than 50 years. When she retired three years ago, she lost her focus, became depressed, and ended up in a hospital and then a rehabilitation facility. Her daughter, Iona, who is 46 and an only child, worried about the next step. Thanks to Home Instead, a national franchise that provides nonmedical caregivers by the hour, the next step was for Azzurra to return to her own apartment.
>
> Initially, a caregiver visited every day to help with meal preparation, monitor her medications, drive her to the grocery store, and accompany her to doctor appointments. Once Azzurra was feeling better, the visits were cut back to three times a week.
>
> Iona is pleased that she has found a way to keep her mother in the garden apartment she loves, close to the university and the local senior center, where she teaches Italian. Iona relies on the home-care agency to act as her eyes and ears. She also has a maid service clean the apartment twice a month, pays $15 a month for an emergency-response bracelet that her mother can use to summon help, and phones every day—even if her mom doesn't always remember the calls. (Franklin, 2000, p. 88)[*]

When the case manager's work focuses on the role of *broker*, the emphasis is on linking the client to other services. Interaction with the client is done prima-

---

[*]Source: This excerpt and the following ones are from "Caring Across the Miles," by M.B. Franklin, 2000, *Kiplinger's Personal Finance Magazine, 54,* pp. 86–90. Reprinted with permission of Kiplinger's.

rily to assess the person and the environment, link him or her to services, and monitor the services delivered by others. The case manager in this role provides fewer direct services than does the generalist. For example, individuals who are responsible for elderly family members or those who are disabled or ill often find it difficult to fulfill their responsibilities when they live in different towns, states, or parts of the country. Case-managers-for-hire is a new trend that helps meet these needs.

> Professional geriatric-care managers, who typically charge from $60 to $200 an hour, can be particularly helpful when you're trying to provide care across the miles. They can assess your parent's condition, acquaint you with local services and residential facilities, and recommend solutions in keeping with your parent's—or your—financial resources. You can get a list of managers at the National Association of Professional Geriatric Care Managers' Web site <www.caremanager.org> or by writing the group at 1604 N. Country Club Rd., Tucson, AZ 85716. (Franklin, 2000, p. 90)

In some cases, the case manager is a counselor or a *therapist*, personally providing this service. Often in the mental health field, the client seeks the help of the counselor for a presenting problem. The counselor presents credentials and areas of expertise and works with the client to determine whether there is a match with the client's problem. Referral elsewhere occurs when there is not a good match, or if the client later experiences a crisis during the helping process.

After the death of her husband, Althea, age 64, was encouraged by her family to see a therapist in private practice. They believed she was suffering from depression. The therapist saw her once a week for five years and three months. At the beginning of treatment, the therapist conducted a mental health assessment using a depression checklist. On occasion, the therapist referred Althea to professionals for help with nutrition, exercise, and part-time employment. Althea and the therapist mutually agreed to conclude the treatment at the end of five years.

In many situations, clients need services because they become physically or mentally ill, sometimes catastrophically so. There is often a case manager responsible for *cost containment*. The case manager must make recommendations after considering outcomes, levels and quality of care, and expense. The case manager gets the client, family, and professionals involved in decisions. Negotiating services to find the best quality for the lowest price is one responsibility of the case manager. He or she also documents the negotiated arrangements, prepares a cost analysis statement, and tracks all interactions between the client and the providers. Box 3.1 lists the information included in a standard cost-containment analysis; Box 3.2 gives an example.

> ### BOX 3.1   Summary of Cost-Containment Analysis Variables
> ● ● ● ● ● ● ● ● ● ● ● ●
>
> 1. Identifiers (file, case, or Social Security number, carrier, employer group; date case opened; date case closed; total weeks or months in case management; diagnosis)
> 2. Overview of case management intervention
> 3. Summary of intervention
> 4. Case management fees
> 5. Savings, such as avoided charges, potential charges, discounts and/or negotiated reductions, reductions in services, products, and equipment
> 6. Actual charges
> 7. Gross savings (potential charges minus actual charges)
> 8. Net savings (gross savings minus case management fees)
> 9. Status of case (open or closed)
>
> SOURCE: From *The Case Manager's Handbook*, by C. Mullahy, pp. 283–284. Copyright © 1998 Aspen Publishers, Inc. Reprinted with permission.

Organization-Based Case Management

In this model, the nature of case management is determined by the organizational structure of the agency or organization. In other words, the way in which services are arranged determines how services are delivered. The three examples of organization-based case management presented here illustrate the collaborative atmosphere that exists among many helping professionals. Each person has a specific assignment and responsibility; the services are organized so that relationships between professionals are integrated to serve the clients' needs better.

Goal   Case managers meet multiple needs through a single point of access, with one location for service delivery. This comprehensive service delivery sometimes resembles that provided by the traditional extended family.

Responsibilities   The organization-based model provides comprehensive case management: Each client receives an individual assessment and plan, which may include social support, housing, recreation, work, and time to integrate into the community. The case manager's responsibilities range from coordinating services (supervision of intake, assessment, planning, brokering, monitoring, and termination) to leading a team of professionals who provide services to the client. Within the second scenario, the case manager role varies: Sometimes there is a professional whose primary responsibility is management of the case, and at other times the one who initiates services also assumes the management role.

Primary Roles   Advocate, broker, collaborator, coordinator, evaluator, expediter, planner, problem solver, recordkeeper, and system modifier.

BOX 3.2   Case Report
●●●●●●●●●●●●

HCMC#: 5430

Group: Company X 123-100

Date of referral: 1/2/XX

Diagnosis: Bronchopulmonary dysplasia

Status: Open

This child remains home vs. hospitalization. R.N. care is provided 16 hours/day × 7 days plus oxygen and medical supplies. Child has expended $100,000 of nursing charges this year on or about September 10.

Referral has been made and approval granted for Care at Home program retroactive to September 1, XXXX. Care at Home will share cost of home care with carrier and family. Referral to and approval for this program was through direct and ongoing intervention by Health Care Management Consultants.

Cost savings are represented by prevention of hospitalization, reduced cost for DME and supplies, and stabilization of condition. Parents' requests for increased services have not been deemed warranted.

Avoidance of potential charges:

Potential hospitalization 90 days × $1,000	$90,000.00
Cost of P.T., O.T., S.T. $50 × 5 treatments/wk × 12 wk	3,000.00
	$93,000.00

Actual charges:

R.N. care 16 hours/day × 7 days $520/day × 90 days	$46,800.00
DME and supplies	15,600.00
	$62,400.00
Case management fees	$ 1,072.00
Total actual charges:	$63,472.00

Net savings (potential charges minus actual charges): $29,528.00

SOURCE: From *The Case Manager's Handbook,* by C. Mullahy, pp. 247–248. Copyright © 1998 Aspen Publishers, Inc. Reprinted with permission.

Length of Involvement   The duration varies. If the case is complicated and several specialists are needed, services are provided longer. In other cases, short-term service is adequate.

Strengths   Services are provided on an inpatient, outpatient, or residential basis, but all in one location. Client assessment is multifaceted, with a holistic approach. The plan is individualized and easily monitored. Staff members function as a team, with a common goal, regular meetings, and a common reporting scheme.

Weaknesses   Resource availability may be a problem if the client needs services not available in the center. Service integration depends on clear organizational structure and lines of authority; the staff must agree on the problem, the plan, and the implementation. Resource availability can be a problem. The family of the client may be less engaged in the helping process than in other models. In addition, the client may become accustomed to the environment and never grow beyond it.

ILLUSTRATIONS

Many case managers see the multi-service center as the ideal. They believe it reduces the risk of people getting lost in the system, clients' feelings of frustration, and duplication of information, forms, and services. Clients can have a single intake interview and just one assessment. They have one case manager who has access to all the professionals involved in the delivery of services.

An example of the use of comprehensive service centers is a new approach to public welfare that has transformed the eligibility worker into a case manager. The ultimate goal is to move families toward self-sufficiency. Case managers collect information to determine eligibility and assess the client's use of other social support services, such as child care, health services, housing employment, and medical benefits. All these services are available in the neighborhood or a short bus ride away.

The case management team members are all employees, managers, supervisors, case managers, specialists, and clerks. Everyone from the same office is trained together in the operations of case management, problem solving, and teamwork, and all are committed to helping clients become self-sufficient.

Another example of the organization-based model is the interdisciplinary team, which is particularly effective for clients with complex problems that require the involvement of several professionals. This approach brings specialists together with the mutual responsibility to help the client. Difficulties may arise when they do not agree, when the case manager or the specialists have too heavy a caseload, or when resources are insufficient for the services needed.

> The Far West Mental Health Service is based in Broken Hill, a remote town of 24,000 people in the far west of New South Wales (NSW). The Far West Mental Health Service covers a geographical area of 147,000 kilometres. Broken Hill is a mining town, although like many parts of rural Australia has high rates of unemployment and is suffering the effects of the rural recession. This semiarid area is 500 kilometres from Adelaide and 1300 kilometres from Sydney. The nearest NSW psychiatric hospital is 900 kilometres away. The service is community-based, with a 24-hour crisis component, a small two-bedded in-patient unit, rehabilitation service and an outreach service, in partnership with the Royal Flying Doctor, to isolated communities. The majority of consumers are treated in the community. The service has a multidisciplinary team of 10 staff, including psychiatric nurses, social workers, and psychologists, who all practice case management. Within the service, as with many other mental health services in Australia, is recognition that

the components of clinical case management represent the core tasks of all mental health clinicians. Discipline-specific skills are used in addition to these core skills. (Hemming & Yellowlees, 1997, pp. 589–590)[*]

Many professionals regard the psychosocial rehabilitative center as an excellent treatment approach. As with other organization-based services, the care is comprehensive, and the client receives social support. Among the criticisms are that these centers may resemble institutions and that they can be very expensive.

John H., a senior at a college in New England, was hit by a car while attending Mardi Gras in New Orleans. After hospital discharge, John entered a rehabilitation center for the assessment of job skills and personal and social adjustment. The brain damage he suffered is permanent, and he must cope with a number of limitations.

The center provides vocational assessment, counseling, independent living skills, and education. Staff also dispense medication and make referrals when necessary.

Responsibility-Based Case Management

The focus of responsibility-based case management is the transition of care from human service professionals to non-professionals. Often clients continue to need assistance long after the professional case managers have terminated their work with clients and families. To meet the on-going or recurring need for case management services, teams of family, friends, or community volunteers are trained to provide continuing case management. Professional case managers support the caretaker and are available to help during emergencies, crises, and other stressful times.

Goal   The responsibility-based model emphasizes both short- and long-term involvement of the case manager, the coordination of services, the help of volunteers, and the empowerment of clients.

Responsibilities   The individual or group responsible for case management provides coordination, finds assessment services, and networks with others in the human service delivery system to provide access to needed specialists and services. Problem identification, plan development, and implementation are other responsibilities. The case manager also provides support and assistance in making and maintaining other linkages.

Primary Roles   Broker, collaborator, coordinator, evaluator, expediter, planner, problem solver, recordkeeper, and system modifier.

Length of Involvement   The involvement may be short-term, during a crisis or developmental problem, or long-term, as with a physical or mental illness, a disability, or geriatric problems.

Strengths   The responsibility-based model allows case management responsibilities to be assumed by various individuals or groups, including family, neighbors, volunteers, and the client. In many cases, the designated case manager may already have an established relationship with the client. His or her involvement helps the client who does not have easy access to services. Under this model, service delivery is cost effective, the community is involved, and independence is encouraged.

Weaknesses   In some cases, the person designated as case manager may not have the client's best interests at heart, may lack the necessary knowledge and skills, or may be ineffective in monitoring service provision. Training and supervision may be costly. Accepting family members or volunteers as case managers may be difficult for the client. Case managers who are not part of the human service delivery system may have trouble coordinating and gaining access to services.

ILLUSTRATIONS

It is a current trend in human services to ask families to act as case managers and then provide them with the support to do so. With costs escalating and institutional care being replaced by community care, it is cost effective for families to perform this central role. For such a system to be effective, the family must receive continuing education about the human service delivery system and must have professional help available when crises arise.

> Shirley Poll of West New York, NJ, has been a long-distance caregiver for her mother and stepfather for five years. Every six weeks or so she flies to East Lansing, MI, to visit them in the nursing home where they now live.
>
> Poll created a five-member team of family, friends, and heath care professionals to help her manage her parents' care. Team members advise her on her parents' health, handle finances and powers of attorney, and are regular visitors. They write comments in a shared diary to keep each other up to date on new developments. Her advice to other long-distance caregivers: Communicate regularly with the people responsible for your parents' hands-on care. (Franklin, 2000, p. 89)

The responsibility-based case management model also applies to supportive care, which is used most often when clients cannot travel to receive services. Most affected are people located in rural areas and those who live alone—often persons who are disabled or elderly. Sometimes the services provided are minimal, but they are helpful to people who otherwise would not be served at all.

An example of a supportive care program is one that uses *gatekeepers*—individuals who happen to have contact with elderly prospective clients in their day-to-day work.

Elderly individuals who live alone are one type of client served by a gatekeeper program. Ms. H. is 82 years old and has been widowed for 15 years. Four weeks ago, she sprained her ankle on her weekly walk to the grocery store and has not left her apartment since. Her apartment manager, realizing that he had not seen her for a while, visited her one afternoon. During the visit, it became clear to him that she was having difficulty caring for herself. She had lost weight and she was unkempt; her apartment was dirty and messy. After the visit, he called the Gatekeeper Program to get assistance for her.

Because of the rising cost of service delivery, many agencies and communities have developed strong volunteer programs. The agencies provide excellent training, on-going education, and good supervision, thus allowing volunteers to assume case management responsibilities. In this way, the volunteers are able to contribute to the welfare of their local community.

Many home health care agencies use volunteers to help people more than 70 years of age remain in their own homes. The project is directed by a team consisting of a nurse, social worker, environmental assessment specialist, volunteer coordinator, and program administrator. Since the case managers for the Department on Aging have caseloads greater than 200 clients, this program supplements the care provided by the Department on Aging.

The day-to-day contact with elderly clients is maintained by volunteers, who are supervised and assisted by community workers—paraprofessionals who themselves live in the neighborhoods. The services the volunteers provide might include grocery shopping, cooking, cleaning, companionship, and making others aware of any special needs or emergencies. One volunteer, a Cuban immigrant, was herself a client in the program when her husband was terminally ill. Now, she uses her language skills to help clients who speak only Spanish.

The responsibility-based case management model is applicable when the client is the case manager. This approach develops the client's maximum potential, which is emphasized by strengths-based case management (Whitley, White, Kelley, & Yorke, 1999). This type of case management stresses building on the strengths and resources of individuals rather than more traditional approaches that focus on deficits and needs. Self-determination—the right to establish one's own goals and to have an active role in problem resolution—is of primary importance here. The belief is that a client who has learned to act as manager can provide long-term care for himself or herself and for others.

With the support of professional case managers, breast cancer survivors may serve as care managers for themselves or for newly diagnosed patients. They complete an in-service training program at the hospital's Breast Center. Although they do not provide medical services, they do counsel, arrange for prosthetic consultations, refer for group discussion and counseling, and talk with family members.

Case Management and the Problem-Solving Process

Problem solving, an integral part of the helping process, can be defined in many ways. For Brill (1998), problem solving is an orderly way of thinking and planning. Egan (1998) views it as a management strategy that can be used to develop opportunities. It is defined by Epstein (1985) as the ability to identify problems and solve them systematically.

Our model for problem solving includes three components: problem identification, decision making, and problem resolution (McClam & Woodside, 1994). The goals of the model are twofold: solving the problem and teaching the problem-solving process. The roles a helper uses in the problem-solving process are data gatherer, broker, teacher, evaluator, caregiver, planner, and advocate. The helper is a data gatherer and a broker during the problem-identification stage. During the decision-making stage, the roles of teacher, evaluator, and caregiver are particularly important. Problem resolution calls for the roles of planner, advocate, evaluator, and teacher.

Much of the work in case management is directly related to problem solving. What is the relationship between these two processes? What are the distinctions between one process and the other?

Problem solving and case management have many similar characteristics. First, both processes can be outlined and described: They are not linear, they can be learned, and both are applied in the helping process. Second, problem solving and case management use many similar roles to reach goals and outcomes, such as data gatherer, broker, and evaluator. Third, both processes are grounded in concepts of helping, such as empowering clients and critically evaluating the delivery of services.

The difference between problem solving and case management is one of scope. Case management can be seen as an umbrella, under which services are provided for clients, especially those with multiple needs. It includes assessment, the development of treatment plans, and the implementation of service plans through a process of providing and coordinating services. Problem solving is subsumed in case management as one of the responsibilities that a case manager might assume and one of the tools that can be used. It is especially appropriate that the case manager use the problem-solving process during the planning phase and the implementation of treatment.

A critical component of both case management and problem solving is creativity: finding innovative ways of approaching difficulties and opening up opportunities. When case management incorporates imagination and resourceful problem solving, case managers and clients are more likely to be satisfied with the process (McClam & Woodside, 1994).

 ## Chapter Summary

Case managers perform many roles that form the basis of their professional responsibilities. Thirteen roles that are integral to case management are discussed in this chapter. A case manager may function primarily in one role and engage in other roles sporadically.

Models of case management contribute to an understanding of case management. They demonstrate that case management is a flexible process; illustrate the different goals that determine roles, responsibilities, and length of client involvement; and identify strengths and weaknesses. The three models are role-based, organization-based, and responsibility-based case management. Role-based case management focuses on case manager roles in service delivery. The configuration of services to meet multiple client needs defines organization-based case management. Family members, a supportive care network, volunteers, and the client are examples of case managers in the responsibility-based model.

Problem solving is another fruitful way of thinking about case management. They are similar in characteristics but different in scope. Problem solving is one role or responsibility in case management.

Chapter Review

◆ Key Terms ....

Advocate	*Planner*
Broker	*Problem solver*
Coordinator	*Recordkeeper*
Collaborator	*System modifier*
Community organizer	*Role-based case management*
Consultant	*Organization-based case management*
Counselor	*Responsibility-based case management*
Evaluator	*Problem solving*
Expediter	

◆ Reviewing the Chapter ....

1. Describe situations in an agency setting in which you might function in the following roles: planner, advocate, broker, expediter, recordkeeper.
2. Why is it important to understand the different models of case management?

3. How are the three models of case management different?
4. Describe a situation in which the role-based model of case management is applicable.
5. How does the helping professional function in the organization-based model?
6. Describe the characteristics of the responsibility-based model.
7. Compare and contrast the roles of case managers and clients in each of the three case management models.
8. What is the relationship between case management and problem solving?

◆ *Questions for Discussion* ....

1. Why do you think so many roles are involved in case management? Cite an example to illustrate your answer.
2. Now that you have completed your reading about the models of case management, what can you conclude about the definition of case management?
3. Suppose you wanted to use case management to deliver services to elderly people in your community. How would you determine which model would apply?
4. If you were working as a case manager in children's protective services, how would you determine which case management roles would be needed?

References

Austin, M. J. (1981). *Supervisory management for the human services.* Englewood Cliffs, NJ: Prentice Hall.

Brill, N. I. (1998). *Working with people: The helping process (6th edition).* White Plains, NY: Longman.

Egan, G. (1998). *The skilled helper: A problem-management approach to helping.* Pacific Grove, CA: Brooks/Cole.

Epstein, L. (1985). *Talking and listening: A guide to the helping interview.* St. Louis: Times Mirror/Mosby.

Franklin, M. B. (2000, November). Caring across the miles. *Kiplinger's Personal Finance Magazine, 54*(11), 86–90.

Hemming, M., & Yellowlees, P. (1997). An evaluation study of clinical case management using clinical case management standards. *Journal of Mental Health, 6*(6), 589–598.

Jackson, B., Finkler, D., & Robinson, C. (1992). A case management system for infants with chronic illnesses and developmental disabilities. *Children's Health Care, 21*(4), 224–232.

McClam, T., & Woodside, M. (1994). *Problem solving in the helping professions.* Pacific Grove, CA: Brooks/Cole.

Mullahy, C. (1998). *The case manager's handbook.* Gaithersburg, MD: Aspen.

Weil, M., & Karls, J. (1985). Key components in providing efficient and effective services. In M. Weil & J. Karls (Eds.), *Case management in human service practice* (pp. 29–71). San Francisco: Jossey-Bass.

Whitley, D. M., White, K. R., Kelley, S. J., & Yorke, B. (1999). Strengths-based case management: The application to grandparents raising grandchildren. *Families in Society, 80*(2), 110–119.

Chapter Four

The Assessment Phase of Case Management

F *irst I get the referral. Then I do a screening on the family to find out if they are suitable for our program. I set up an appointment to make a home visit. Sometimes they come into the office first and the home visit is made afterward. I try first to verify their HIV status through their medical providers or clinic doctor. Then we do an assessment of the family to identify their needs and strengths. I follow up with the referral source to find out if all the needs are being met. If not, I make sure that things are put in place.*

> —Carolyn Brown, Third Avenue Family Service Center, Bronx, NY, personal communication, May 5, 1994

W *hen we get a hotline call, basically all we know about this kid is what is said over the phone. . . . There is no way for us to verify the story except with their permission and that is obtained only when they come to the shelter. Trying to feel somebody out over the phone and get as much information as possible is difficult. We need to determine if they are appropriate for the shelters, if they are violent, and if they are a danger to the other kids.*

> —Christine Gossmeyer, Youth in Need, St. Louis, MO, personal communication, December 13, 1999

W *e have a nightly emergency shelter where men can come for a change of clothes and a shower. It is during our chapel time, and our message is that you can change, there is hope for you. We have referrals from other agencies to come here and we have self-referrals. If a person is interested and they want to change or if they are just tired of being what they are, then a person petitions to come into the program. We let them stay here for one week. During that time, we give an initial assessment to see where they are and what problems they have. It's also a time for them to get healthy. But you really can't talk to somebody, you can't do an in-depth assessment because their mind is clouded with drugs or alcohol.*

> —Oliver Ramirez, Miami Rescue Mission, Miami, FL, personal communication, May 13, 1997

Assessment is defined as "the appraisal of a situation and the people involved in it" (Brill, 1998). As the initial stage in case management, assessment generally focuses on identifying the problem and the resources needed to resolve it. Focusing on the people who are involved includes attention to client strengths that can be a valuable resource to encourage client participation and facilitate problem solving. Specifically, the case manager identifies the initial presenting problem and makes an eligibility determination. As the three opening quotations show, data are gathered and assessed at this phase so as to show the applicant's problem in relation to the agency's priorities. Identifying possible actions and services and determining who will handle the case are also part of the assessment phase. These activities occur differently at each agency. Carolyn Brown at the Third Avenue Family Service Center does a preliminary screening after she has received a referral. At the Miami Rescue Mission, the initial contact is a preapplication or intake interview. At Youth in Need, the initial contact is a hotline call.

This chapter explores the assessment stage of case management—the initial contact with an applicant for assistance, the interview as a critical component in data gathering, and the case record documentation that is required during this phase. The assessment phase concludes with the evaluation of the application for services. For each section of the chapter, you should be able to accomplish the following objectives.

APPLICATION FOR SERVICES
- List the ways in which potential clients learn about available services.
- Compare the roles of the case manager and the applicant in the interview process.
- Define *interview*.
- Distinguish between structured and unstructured interviews.
- State the general guidelines for confidentiality.
- Define the case manager's role in evaluating the application.
- List the two questions that guide assessment of the collected information.

CASE ASSIGNMENT
- Compare the three scenarios of case assignment.

DOCUMENTATION AND REPORT WRITING
- Distinguish between process recording and summary recording.
- List the content areas of an intake summary.
- State the reasons for case or staff notes.

 ## Application for Services

Potential clients or **applicants** learn about available services in a number of ways. Frequently, they apply for services only after trying other options. People having problems usually try informal help first; it is human nature to ask for help from family, friends, parents, and children. Some people even feel comfortable sharing their problems with strangers waiting in line with them or sitting beside them. A familiar physician or pastor might also be consulted on an informal basis. On the other hand, some people avoid seeking informal help because of embarrassment, loss of face, or fear of disappointment.

Previous experiences with helping agencies and organizations also influence the individual's decision to seek help. Many clients have had positive experiences with human service agencies, resulting in improved living conditions, increased self-confidence, the acquisition of new skills, and the resolution of interpersonal difficulties. Others have had experiences that were not so positive, having encountered helpers who had different expectations of the helping process, delivered unwanted advice, lacked the skills needed to assist them, were inaccessible, or never understood their problems. Increasingly, clients may also encounter local, state, and national policies that may make it difficult to get services. An individual's prior experiences also play a role in the decision regarding whether to seek help.

An individual who does decide that help is desirable can find information about available services from a number of sources. Informal networks are probably the best sources of information. Family, friends, neighbors, acquaintances, and fellow employees who have had similar problems (or who know someone who has) are people we trust to tell us the truth about seeking help. Other sources of information are professionals with whom the individual is already working, the media (posters, public service announcements, and advertisements), and the telephone book. Once people locate a service that seems right, they generally get in touch on their own (self-referral).

Other individuals may be referred by a human service professional if they are already involved with a human service organization but need services of another kind. They may be working with a professional, such as a physician or minister, who also makes referrals. These applicants may come willingly and be motivated to do something about their situation, or they may come involuntarily because they have been told to do so or are required to do so. The most common referral sources for mandated services are courts, schools, prisons, protective services, marriage counselors, and the juvenile justice system. These individuals may appear at the agency but ask for nothing, even denying that a problem exists.

The words of case managers at four different locations illustrate the various ways referral can happen. The first speaker works at a community-based agency.

> St. Patrick Center is open for referrals five days a week, 52 weeks of the year. We have a Mobile Outreach identify clients. A van or bicycles staffed by two counselors is on the streets of the city of St. Louis making contact with people who are homeless, offering them assistance and services. Typically 200+ clients come per day for services. We know the length of time they stay in our program, some stay a month or so . . . sometimes as long as a year. (Jan Rasmussen, St. Patrick Center, St. Louis, MO, personal communication, March 6, 2002)

The second quote is from a caseworker at an institution that serves the elderly in a large metropolitan city.

> We get clients through referrals—word of mouth, professionals, people from Discharge Planning, agencies that serve the visually impaired, etc. We did do outreach in the community, but that hasn't seemed to net us the most clients. If I look through my book, I would find that most of our referrals in the past have come from the Jewish Guild for the

Blind. Among the city agencies, the access to services has a little more continuity to it. If certain people need certain services, they are required by law to pass them to that agency. If someone in the Medicaid system needs a certain type of care, they must be referred to long-term care. (Rosalyn Baum, personal communication, May 3, 1994)

The third speaker works at a community mental health center that coordinates services for clients with chronic mental illness. At this agency, client records are available at referral that provide important information about the client and previous services.

We will get a referral. I will get the information and research it. Does the client on paper look like he or she meets our criteria? If they do, I or one of the other case managers will go to wherever the client is, usually in the hospital. We like to get them when they are in the hospital because they are contained. It is easier to find them. Plus if we decide to pick them up and we meet them in the hospital, they know who we are when we come to see them in the community. Nine times out of ten, if we have not made contact in the hospital, they will not open the door to us. They do not know who we are. We interview them, talk about our program, and why they were referred to us. (Jan Cabrera, personal communication, March 24, 1998)

The final example is from a large hospital.

Our committee, which is interdisciplinary—recreation, social work, nursing, and dietary—discusses the referral to see if we can meet some of that person's needs, and if that person can be helped in the program. Sometimes there is a question whether he or she can. We then decide if we can give it a try. The person may not have sufficient medical problems or service needs to be eligible for our program. It is our obligation to refer people to the appropriate level of care. (Roz Jaffe, personal communication, May 3, 1994)

These examples illustrate the different ways in which a referral occurs. In an institutional setting, referrals are made by other professionals and departments in the institution. In other settings, referrals can also occur in-house or come from other agencies or institutions. Sometimes the referral procedure may include an initial screening by phone (Figure 4.1), a committee deliberation, an individual interview, or the perusal of existing records or reports, or both. Usually, the individual who receives the referral makes sure that the necessary paperwork is included.

Not all applicants who seek help or are referred for services become clients of the agency. **Clients** are those who meet eligibility criteria to be accepted for services. An intake interview is the first step in determining eligibility and appropriateness.

PROJECT LIVE INTAKE FORM

DATE _____

NAME _____ AGE _____

PHONE _____

ADDRESS AND DIRECTIONS _____

_____

_____

REFERRAL SOURCE (AND REASON OR RELATIONSHIP) _____

_____

PROBLEM _____

_____

_____

OTHER _____

Figure 4.1   Phone screening form

The Interview

An **interview** is usually the first contact between a helper and an applicant for services, although some initial contacts are by telephone or letter. The first helper an applicant talks with may be an intake worker who only conducts the initial meeting, or he or she may in fact be the case manager or service provider. If the first contact is an intake worker, the applicant (if accepted for services) will be assigned to a case manager, who will coordinate whatever services are provided. In this text, we will assume that the intake interviewer is the case manager.

The initial meeting with an applicant takes place as soon as possible after the referral. The interview is an opportunity for the case manager and applicant to get to know one another, define the person's need or problem, and give some structure to the helping relationship. These activities provide information that becomes a starting point for service delivery, so it is important for the case manager to be a skillful listener, interpreter, and questioner (skills that you will read about in Chapter 5). First, we will explore what an interview is, the flow of the interview process, and two ways to think about interviews that you may conduct as part of the intake process.

Interviewing is a critical tool for communicating with clients, collecting information, determining eligibility, and developing and implementing service plans—in all, a key part of the case management process. Primary objectives of interviewing are to help people explore their situation, to increase their understanding of it, and to identify client resources and strengths. The roles of the applicant and the case manager during this initial encounter reflect these objectives. The applicant learns about the agency, its purposes and services, and how they relate to his or her situation. The case manager obtains the applicant's statement of the problem and explains the agency and its services. Once there is an understanding of the problem and the services the agency offers, the case manager confirms the applicant's desire for services. The case manager is also responsible for recording information, identifying the next steps in the case management process, informing the applicant about eligibility requirements, and clarifying what the agency can legally provide a client. There are three desirable outcomes of the initial interview (McGowan & Porter, 1967). First, the applicant feels free to express himself or herself. Second, the applicant leaves confident of being able to work with the case manager toward a satisfactory solution. Finally, rapport is established between the two participants.

Exactly what is an interview?   In case management, an interview is usually a face-to-face meeting between the case manager and the applicant; it may have a number of purposes, including getting or giving information, resolving a disagreement, or considering a joint undertaking. Another description is a directed conversation, "an event composed of a sequence of physical and mental experiences that occur when and where a helping professional practitioner and a client talk to one another" (Epstein, 1985, p. 2). It has also been described as "professional conversation" that "involves communication between two people" (Garrett, 1972, p. 5).

An interview may also be an assessment procedure. It can be a testing tool in areas such as counseling, school psychology, social work, legal matters, and employment applications. One example of this is testing an applicant's mental status (discussed further in Chapter 6). We can also think of the interview as an assessment procedure in which one of the first tasks is to determine why the person is seeking help. An assessment helps define the problem, and the resulting definition then becomes the focus for intervention.

Another way to think about defining the interview is to consider its content and process (Enelow & Wexler, 1966). The *content* of the interview is what is said, and the *process* is how it is said. Analyzing an interview in these terms gives a systematic way of organizing the information that is revealed in the interaction between the case manager and the applicant. It also facilitates an understanding of the overall picture of the individual. This is particularly relevant in case management, since many clients have multiple problems. You will read more about process and content in the next chapter.

Ivey and Ivey (1999) caution that although the terms *counseling* and *interviewing* are sometimes used interchangeably, interviewing is considered the more basic process for information gathering, problem solving, and the giving of information or advice. An interview may be conducted by most anyone—business people, medical staff, guidance personnel, or employment counselors. Counseling is a more intensive and personal process, often associated with professional fields such as social work, guidance, psychology, and pastoral counseling. Interviewing is a responsibility assumed by most case managers, whereas counseling is not always the job of the case manager.

How long is an interview?   An interview may occur once or repeatedly, over long or short periods of time. However, Okun (2002) limits the use of the word *interview* to the first meeting, calling subsequent meetings *sessions*. In fact, the actual length of time of the initial meeting depends on a number of factors, including the structure of the agency, the comprehensiveness of the services, the number of people applying for services (an individual or a family), and the amount of information needed to determine eligibility or appropriateness for agency services.

A case manager from a community action agency recalled that she had conducted an intake interview in 15 minutes. "When I was in school reading textbooks, I thought I would always have an hour and a half for every interview. And we would just ask questions to get all the information about this person. Not true!" (Janelle Stueck, Private Industry Council, Knoxville, TN, personal communication, August 30, 1993).

Where does the interview take place?   Interviews generally take place in an office at agencies, schools, hospitals, and other institutions. Sometimes, however, they are held in an applicant's home. In such cases, the case manager has the distinct advantage of observing the applicant in the home, which gives information about the applicant that may not be available in an office setting. An informal location, such as a park, a restaurant, or even the street, can also serve as the scene of an interview. Whatever the setting, it is an important influence on the course of the interview (Hutchins & Cole Vaught, 1997).

What do all helping interviews have in common?   There are commonalities to all interviews in the human services that should occur in any initial interview. First, there must be shared or mutual interaction: Communication between the two participants is established, and both share information. The case manager may be sharing information about the agency and its services, while the applicant may be describing the problem. No matter what the subject of their conversation is, the two participants are clearly engaged as they develop a relationship.

A second factor is that the participants in the interview are interdependent and influence each other. Each comes to the interaction with attitudes, values, beliefs, and experiences. The case manager also brings the knowledge and skills of helping, while the person seeking help brings the problem that is causing distress. As the relationship develops, whatever one participant says or feels triggers a response in the other participant, who then shares that response. This type of exchange builds the relationship through the sharing of information, feelings, and reactions.

The third factor is the interviewing skill of the case manager. He or she remains in control of the interaction and clearly sets the tone for what is taking place. Perlman (1979) stresses the helper's knowledge and skills as a characteristic of the helping relationship. The knowledge and expertise of the case management process distinguishes the case manager from the applicant and from any informal helpers who have previously been consulted. Since the helping relationship develops for a specific purpose and often has time constraints, it is important for the case manager to bring these characteristics to the interaction, in addition to providing information about the agency, its services, the eligibility criteria, community resources, and so forth.

THE INTERVIEW PROCESS

The interview's structure refers to the arrangement of its three parts: the beginning, the middle, and the end. The beginning is a time to establish a common understanding between the case manager and the applicant. The middle phase continues this process, through sharing and considering feelings, behaviors, events, and strengths. At the end, a summary provides closure by describing what has taken place during the interview and identifying what will follow. Let's examine each of these parts in more detail.

The beginning   Several important activities occur at the beginning of the interview: greeting the client, establishing the focus by discussing the purpose, clarifying roles, and exploring the problem that has precipitated the application for services. The beginning is also an opportunity to respond to any questions that the applicant may have about the agency and its services and policies. The following questions are those most frequently raised by clients (Weinrach, 1987).

- How often will I come to see you?
- Can I reach you after the agency closes?
- What happens if I forget an appointment?
- Is what I tell you confidential?
- What if I have an emergency?
- How will I know when our work is finished?
- What will I be charged for services?
- Will my insurance company reimburse me?

Answering these questions can lead to a discussion of the applicant's role and his or her expectations for case management.

The middle   The next phase of the interview is devoted to developing the focus of the relationship between the case manager and the applicant. Assessment, planning, and implementation also take place at this time. Assessment occurs as the problem is defined in accordance with the guidelines of the agency. Often assessment tends to be problem focused, but a discussion that focuses only on the client's needs or weaknesses, or both, can be depressing and discouraging. Spending some of the interview identifying strengths can be energizing and result in a feeling of control. Assessment also includes consideration of the applicant's eligibility for services in light of the information that is collected. All these activities lead to initial planning and implementation of subsequent steps, which may include additional data gathering or a follow-up appointment.

The end   At the close, the case manager and the applicant have an opportunity to summarize what has occurred during the initial meeting. The summary of the interview brings this first contact to closure. Closure may take various forms, including the following scenarios. (1) The applicant may choose not to continue with the application for services. (2) The problem and the services provided by the agency are compatible, the applicant desires services, and the case manager moves forward with the next steps. (3) The fit between the agency and the applicant is not clear, so it is necessary to gather additional information before the applicant is accepted as a client.

A technique that some case managers find successful as an end to the initial interview is a homework assignment that once again turns the applicant's attention to strengths. Individuals may be asked to complete a self-assessment instrument such as the Strengths Self-Assessment Questionnaire, a 40-question self-report instrument designed to involve the client and significant others in identifying strengths (McQuaide & Ehrenreich, 1997). Although this type of questionnaire can be administered at any time in the case management process, the value in such a technique at this time is the movement that might occur from problem-oriented vulnerability to problem-solving resilience. It may also solidify the client's intent to return and promote his or her involvement in the case management process.

STRUCTURED AND UNSTRUCTURED INTERVIEWS

Interviews may be classified as structured or unstructured. A brief overview of these two types of interviews is given here; the next chapter will provide more in-depth information about the structured interview in connection with the discussion of skills for intake interviewing.

Structured interviews are directive and focused; they are usually guided by a form or a set of questions that elicit specific information. The purpose is to develop a brief overview of the problem and the context within which it is occurring. It can range from a simple list of questions to soliciting an entire case history.

Agencies often have application forms that applicants complete before the interview. If the forms are completed with the help of the case manager during the interview, the interaction is classified as a structured one. Of course, this is a good way to establish rapport and to identify strengths, but case managers must be cautious. The interview can easily become a mere question-and-answer session if it is structured exclusively around a questionnaire. Asking yes/no questions and strictly factual questions limits the applicant's input and hinders rapport.

The intake interview and the mental status examination are two types of structured interviews, each of which has standard procedures. Generally, an agency's **intake interview** is guided by a set of questions, usually in the form of an application. The **mental status examination** (which typically takes place in psychiatric settings) consists of questions designed to evaluate the person's current mental status by considering factors such as appearance, behavior, and general intellectual processes. In both situations, the case manager has responsibility for the direction and course of the interview, even though the areas to be covered are predetermined.

In contrast, the **unstructured interview** consists of a sequence of questions that follow from what has been said. This type of interview can be described as broad and unrestricted. The applicant determines the direction of the interaction, while the case manager focuses on giving reflective responses that encourage the eliciting of information. Helpers who use the unstructured interview are primarily concerned with establishing rapport during the initial conversation.

CONFIDENTIALITY

Our discussion of the initial meeting would be incomplete if we did not address the issue of **confidentiality**. Human service agencies have procedures for handling the records of applicants and clients and for maintaining confidentiality; all case managers should be familiar with them. An all-important consideration is access to information.

Every communication during an interview should be confidential in order to encourage the trust that is necessary for the sharing of information. Generally, human service agencies allow the sharing of information with supervisors, consultants, and other staff who are working with the applicant. The client's signed consent is needed if information is to be shared with staff employed by other organizations. The exception to these general guidelines is that information may be shared without consent in cases of emergency, such as suicide, homicide, or other life-threatening situations.

Applicants are frequently concerned about who has access to their records. In fact, they may wonder whether they themselves do. Legally, an individual does indeed have access to his or her record; the Federal Privacy Act of 1974 established principles to safeguard clients' rights. In addition to the right to see their records, clients have the right to correct or amend the records.

There are two potential problem areas related to confidentiality. First, it is sometimes difficult in large agencies to limit access of information to authorized staff. Support staff, visitors, and delivery people come and go, making it essential that records are secure and conversations confidential (Hutchins & Cole Vaught,

1997). The second problem has to do with **privileged communication**, a legal concept under which clients' "privileged" communications with professionals may not be used in court without client consent (Corey, Corey, & Callanan, 1998). State laws determine which professionals' communications are privileged; in most states, human service workers are not usually included. Thus, the helper may be compelled to present in court any communication from the applicant or client. Sometimes it can be a challenge for the case manager to explain this limitation to an applicant while trying to gather essential information. It can also be perplexing to the applicant.

Evaluating the Application for Services

During this phase, the case manager's role is to gather and assess information. In fact, this process may actually start before the initial meeting with the applicant, when the first report or telephone call is received, and continue through and beyond the initial meeting. The initial focus on information gathering and assessment then narrows to problem identification and the determination of eligibility for services. This process is influenced to some extent by guidelines and parameters established by the agency or by federal or state legislation. At this point in the process, the case manager must pause to review the information gathered for assessment purposes.

Part of the assessment of available information is responding to the following questions:

- Is the client eligible for services?
- What problems are identified?
- Are services or resources available that relate to the problems identified?
- Will the agency's involvement help the client reach the objectives and goals that have been established?

Reviewing these questions helps the case manager determine the next steps. To answer the questions and evaluate the application for services, the case manager engages in two activities: a review of information gathering and an assessment of the information.

Review of information gathering   Usually the individual who applies for services is the primary source of information. During the initial meeting, the case manager forms impressions of the applicant. The problem is defined, and judgments are made about its seriousness—its intensity, frequency, and duration (Hutchins & Cole Vaught, 1997). As the case manager reviews the case, he or she considers these impressions in conjunction with the application for services, the case notes summarizing the initial contact with the client, and any case notes that report subsequent contacts. The case manager learns more about the applicant's reasons for applying for services, his or her background, strengths and weaknesses, the problem that is causing difficulty, and what the applicant wants to have happen as a result of service delivery. The case manager also uses information and impressions from other contacts. Other information in the file that may contribute to an understanding of the applicant's situation comes from secondary sources, such as the referral source, the client's family, school officials, or an employer. Information from secondary sources that can be part of the case file might be medical reports, school records, a social history, and a record of services that have previously been provided to the client.

An important part of the review of information gathering is to ascertain that all necessary forms, including releases, have been completed and signatures obtained where needed. It is also a good idea at this point to make sure that all necessary supervisor and agency reviews have occurred and are documented.

Assessing information   Once the case manager has reviewed all the information that has been gathered, the information is assessed. Many case managers have likened this part of case management to a puzzle. Each piece of information is part of the puzzle; as each piece is revealed and placed in the file, the picture of the applicant and the problem becomes more complete. Margaret Mikol at SKIP of New York described it as "a large body of information from all kinds of different people, then you have to sort through that and say, 'Now what are the real issues here?' in a concrete fashion." She added that workers "need to gather a large quantity of information and break that down, and be tremendously proactive" (personal communication, May 4, 1994). As a social worker in Houston suggests,

> Part of the assessment is the case manager's ability to see or feel beyond what is going on and to understand it. For example, there are enclaves of witchcraft in this area, a poor white community. There are many different cultures and needs here. If you work here, you sort of become a part of it. You have to know what you are seeing. A person is not necessarily crazy if they just need to go see someone who understands the shaman and the spells. And sometimes the family just needs a more expensive shaman to get rid of the spell. I have seen some profoundly schizophrenic-looking behavior where people say they are hearing voices coming from the television telling them to kill themselves. This was actually a spell that was cleared up as soon as another shaman said the spell was gone. (Walter Rae, personal communication, December 16, 1999)

Knowing what information is in the file, the case manager's task shifts to assessing the information. Two questions guide this activity. Is there sufficient information to establish eligibility? Is additional information necessary? In addition to answering these questions, the case manager also evaluates the information in the file, looking for inconsistencies, incompleteness, and unanswered questions that have arisen as a result of the review.

Is there sufficient information to establish eligibility?   To answer the question of sufficient information, the case manager must examine the available data to determine what is relevant to the determination of eligibility. The quantity of data gathered is less important than its relevance. Human service organizations usually have specific criteria that must be met to find an applicant eligible. The data must correspond to these criteria if the applicant is to be accepted for services.

The criteria for acceptance as a client for vocational rehabilitation services is a good example. Vocational rehabilitation is a state and federal program whose mission is to provide services to people with disabilities so as to enable them to become productive, contributing members of society. Essentially, the criteria for acceptance for services are the following: The individual must have a documented physical or mental disability that is a substantial handicap to employment, and there must be a reasonable expectation that vocational rehabilitation services will render the applicant fit for gainful employment.

During the assessment phase, a vocational rehabilitation counselor assesses the information gathered to determine whether the applicant has a documented disability. The next step is to document that the disability is a handicap to employment. Does the disability prevent the applicant from returning to work? Or, if the person has not been employed, does the disability prevent him or her from getting or keeping a job? If the answers are yes, the counselor's final task is to find support for a reasonable expectation that, as a result of receiving services, the applicant can be gainfully employed. This brings us to the second main question in the information assessment activity.

Is additional information necessary in order to determine eligibility for services? If the answer is no, the case manager and the client are ready to move to the next phase of case management. If the answer is yes—that is, additional information is necessary to establish eligibility—a decision is then made about what is needed and how to obtain it. In the vocational rehabilitation example, the counselor examines the file for the documentation of a disability. Specifically, the counselor is looking for a medical report from a physician or specialist that will establish a physical disability or a psychological or psychiatric evaluation that will establish a mental disability. If the needed information is not in the file, the helper must make arrangements to obtain the necessary reports.

Establishment of eligibility criteria is not the sole purpose of this phase. Data gathered at this time may prove helpful in the formulation of a service plan. Certainly the case manager does not want to discard any information at this point; neither does he or she want to leave unresolved any conflicts or inconsistencies. Relevant, accurate information is an important part of the development of a plan. Ensuring relevance and accuracy at this point in the process saves time and effort and allows the helper and the client to move forward without delay.

 ## Case Assignment

Once eligibility has been established and the applicant accepted for services, there are three possible scenarios, depending on the particular agency. In all three, the applicant becomes a client who is assigned a case manager to coordinate services.

In many instances, the case manager is the same person who handled the intake interview and determination of eligibility. In some agencies, however, there are staff members whose primary responsibility is conducting the intake interview. After a review, the case is then assigned to a case manager—the helping professional who assumes primary responsibility for the case and is accountable for the services given to the client, whether provided personally, by other professionals at the agency, or by helpers at a different agency.

A second scenario involves the *specialized worker*, a term that may refer to either level specialization or task specialization. Level specialization has to do with the overall complexity and orientation of the client and the presenting problem. Is the case under consideration simple or complex? Is it a case of simply providing requested information, or is it a multiproblem situation? Task specialization focuses on the functions needed to facilitate problem resolution. Does the case require highly skilled counseling, or is coordination sufficient?

The third scenario occurs most often in institutions where a team of professionals is responsible for a number of clients. For example, in a facility for children who are mentally retarded, clients interact daily with a staff that includes a teacher, a nurse, an activity coordinator, a cottage parent, and a social worker. These professionals work together as a team to provide services to each client.

 ## Documentation and Report Writing

Documentation and report writing play critical roles in the assessment phase of service coordination. The main responsibilities facing the case manager in this phase are identifying the problem or problems and determining the applicant's eligibility for services. Documentation of these two responsibilities takes the form of **intake summaries** and **staff notes**. Most agencies have guidelines for the documentation of information gathered and decisions made. In this section, we discuss the forms of documentation, their purposes, and how to write them. Before we discuss intake summaries and staff notes, let's distinguish between process recording and summary recording.

Process Recording and Summary Recording

A **process recording** is a narrative telling of an interaction with another individual. In the assessment phase of case management, a process recording shows what each participant has said by an accurate account of the verbal exchange, a factual description of any action or nonverbal behavior, and the interviewer's analysis and observations. The person making the recording should imagine that a tape recorder and a camera are taking in everything that is heard or seen. Of course, since records are required to be brief and goal oriented, the helper would not attempt an exhaustive description, but this approach helps focus the recorder's attention on accuracy and impartiality.

We believe, as does Wilson (1980), that process recording is a useful tool for helping professionals in training, especially those who are learning to be case managers. Tape recorders, video cameras, and VCRs are readily available today, but many agencies and organizations don't have them, and case managers may not have time or authorization to use the equipment. Process recording is still an effective way to hone one's skills of direct observation.

Process recording is most often used with one-on-one interviews. It includes the following elements.

- Identifying information: names, date, location, client's case number or identifying number, and the purpose of the interview.

Paulette Maloney saw the client, Rosa Knight, for the first time on Monday, November 5 at the agency. Ms. Knight is applying for services, and the purpose of the interview was to complete the application form and inform her of the agency's services.

- Observations: description of physical and emotional climate, any activity occurring during the interaction, and the client's nonverbal behavior.

Ms. Knight appeared in my office on time, dressed neatly in a navy dress. I asked Ms. Knight to come in, introduced myself, shook her hand, and asked her to sit down. Although there was little eye contact, she smiled shyly with her head lowered. In a soft voice, she asked me to call her Rosa. Then she waited for me to speak.

- Content: an account of what was said by each participant. Quotes are helpful here, to the extent that they can be remembered.

I explained to Rosa that the agency provides services to mothers who are single parents, have no job skills, and have children who are under the age of 6. She replied, "I am a single parent with two sons who are 18 months old and 3 years old. I have worked briefly as a domestic." I asked her how long she had worked and where. She replied that she worked about 5 months for a woman a neighbor knew. She quit "because the woman was always canceling at the last minute." She related that one time when she went to work, the house was locked up and the family was out of town. During this exchange, Rosa clasped her hands in her lap and looked up.

- Recorder's feelings and reactions: Sometimes this is called a *self-interview*, meaning that the recorder writes down feelings about and reactions to what is taking place in the interview.

I was angry about the way Rosa had been treated in her work situation. She appears to be a well-mannered, motivated young woman who genuinely wants a job so that she can be self-supporting. Her shyness may prevent her from asserting herself when she needs to. Her goal is to get out of the housing project. I wonder if my impressions are right.

- Impressions: Here the recorder gives personal impressions of the client, the problem, the interview, and so forth. It may also be appropriate to make a comment about the next step in the process.

Rosa Knight is a 23-year-old single mother of two sons, ages 18 months and 3 years. She has worked previously as a domestic. Particular strengths seem to be motivation and reliability. Her goal is to receive secretarial training so that she can be self-supporting and move from the projects. Based on the information she has provided during this interview, she is eligible for services. I will present her case at the next staff meeting to review her eligibility.

This type of recording is particularly useful in learning interviewing skills. Austin (1981) suggests that as students are conducting an interview, they should use legal-size paper divided into three columns: Supervisory Comments, Content—Dialogue, and Gut-Level Feelings. This format is beneficial both to supervisors and to those who are learning how to interview. First, the supervisor can make comments next to the interaction or feelings recorded. This format also helps the instructor or supervisor identify areas that have been mastered as well as those requiring additional practice. Second, writing everything that is said and everything that happens provides a complete record for the recorder. Initially, the recorder may leave out something that is considered relatively unimportant but may actually be a critical piece of information. The description should include as much information as the recorder can remember. Finally, recording how one felt as something was being said or happening increases self-awareness and helps the learner differentiate among facts, feelings, and impressions.

The other style of recording, **summary recording**, is preferred in most human service agencies. It is a condensation of what happened, an organized presentation of facts. It may take the form of an intake summary or staff notes (both discussed later in this section). Summary recording is also used for other types of reports and documentation. For example, a diagnostic summary presents case information, assesses what is known about the client, and makes recommendations. (See the Report for Juvenile Court in Chapter 7.) A second example is *problem-oriented recording*, which identifies problems and treatment goals. This type of recording is common in an interdisciplinary setting or one with a team structure. (An example is the Psychological Evaluation, also in Chapter 7.)

Summary recording differs from process recording in several ways. First, a summary recording gives a concise presentation of the interview content rather than an extensive account of what was said. The focus remains on the client, excluding the case manager's feelings about what transpired. Summary recordings usually contain a summary section, which is the appropriate place for the writer's own analysis. Finally, summary recording is organized by topic rather than chronologically. The case manager must decide what to include and omit under the various headings: Identifying Information, Presenting Problem, Interview Content, Summary, and Diagnostic Impressions.

Summary recording is less time-consuming to write, as well as easier to read. It is preferred for these reasons, not to mention the fact that it uses less paper, thereby reducing storage problems. Where computerized information systems are used, a standard format makes information easy to store, retrieve, and share with others. A reminder is in order here: Agencies often have their own formats and guidelines for report writing and documentation. Generally speaking, however, the basic information presented here applies across agency settings.

Intake Summaries

An intake summary is written at some point during the assessment phase. It is usually prepared following an agency's first contact with an applicant, but it may also be written at the close of the assessment phase. For purposes of illustration, assume that it is written after the intake interview. After the case manager conducts the intake interview, he or she assesses what was learned and observed about the applicant. This assessment takes into account the information provided by the applicant, forms that were completed, and any available information about the presenting problem. Client strengths are also identified at this point. The case manager also considers any inconsistencies or missing information. While integrating the information, the case manager also considers the questions that are presented in the previous section. Is the applicant eligible for services? What problems are identified? Are services or resources available that relate to the problems identified? Will the agency's involvement improve the situation for the applicant?

This information is organized into an intake summary, which usually includes the following data.

- Worker's name, date of contact, date of summary
- Applicant's demographic data—name, address, phone number, agency applicant number
- Sources of information during the intake interview
- Presenting problem
- Summary of background and social history related to the problem
- Previous contact with the agency
- Diagnostic summary statement
- Treatment recommendations

Figure 4.2 is a sample intake summary from a treatment program for adult women who are chemically dependent. To qualify for the residential program, applicants must have a child who is 3 years of age or less and has been exposed to

INTAKE SUMMARY
OAK HILL CENTER

CLIENT'S NAME Katie Dunlap   ADMISSION DATE _____

INTERVIEWER _____   DATE 8-11-XX

Katie Dunlap was admitted to Oak Hill Center August 11, XXXX. The client's date of birth is May 22, XXXX. She is from Nashville. This is her first admission to this center. Prior to admission the client resided with her sister at 1010 Western Avenue in Memphis, TN. Her drug of choice is crack cocaine. The client started using at age 14 and the date of last use was August 9, XXXX. She has been in four rehabilitation centers; however, she has not graduated from any of them. The client has 2 daughters: Candy, who is 4 years of age, and Kristy, who is 2. They have been in foster care for one year. The younger child was chemically exposed. The children will be returned to the client one month from the admission date. The client's father and mother were also chemically dependent. Her father died when she was 11. The client stated that her mother had mental health problems and that they do not have a good relationship. The client has not been treated for being emotionally abused by her mother and physically abused by the father of her children. Currently she has no means of income, nor does she have health insurance. A psychological evaluation is being completed at present. She stated that she views herself as happy, easygoing, and appreciative.

_____   _____
INTAKE WORKER'S SIGNATURE   DATE

Figure 4.2 Intake summary

drugs. The day program is available to mothers who have a child older than 3 years. The applicant whose intake is summarized in Figure 4.2 is applying to the residential program, which lasts one year. During this time children live with their parent in one of ten agency apartments. To graduate from the program, the client must be employed or enrolled in school and be free of substance abuse.

Some individuals enter the program voluntarily, and others are ordered to come as part of their probation. On admission, a staff member conducts a 30-minute interview, which is written up and placed in the client's file within two days. The client then receives an orientation to the program and is assigned a care coordinator in charge of that case. The care coordinator and the client then have a more extensive interview, lasting approximately 90 minutes.

Staff Notes

Staff notes, sometimes called *case notes*, are written at the time of each visit, contact, or interaction that any helping professional has with a client. Staff notes usually appear in a client's file in chronological order. They are important for a number of reasons.

- Confirming a specific service. The helper wrote, "John missed work 10 of 15 days this month. He reported that he could not get out of bed. I referred him to our agency physician for a medication review."
- Connecting a service to a key issue. The helper may write, "I observed the client's interactions with peers during lunch," or "I questioned the client about his role in the fight this morning."
- Recording the client's response. "Mrs. Jones avoided eye contact when asked about her relationship with her family," "Janis enthusiastically received the staff's recommendation for job training," or "Joe resisted the suggestion that perhaps he could make a difference."
- Describing client status. Case notes that describe client status use adjectives and observable behaviors: "Jim worked at the sorting task for 15 minutes without talking," and "Joe's parents were on time for the appointment and openly expressed their feelings about his latest arrest by saying they were angry."
- Providing direction for ongoing treatment. Documenting what has occurred or how a client has reacted to something can give direction to any treatment. "During our session today, Mrs. Jones said she felt angry and guilty about her husband's illness in addition to the feelings of sadness she expressed at our last meeting. These feelings will be the focus of our next meeting."

The format of case notes depends on the particular agency, but they are always important. For instance, one substance abuse treatment facility uses a copy of its form for each client every day. A worker on each of the three shifts checks the behavior observed and makes chronological case notes on the other side. These notes allow the case manager and others working with a client to stay up to date on treatment and progress and provide the means of monitoring the case.

Another format for case notes is illustrated by the notes presented below, from a residential treatment facility for emotionally disturbed adolescents. Recorded January 22, 2001, these notes are in the file of a 15-year-old female client who is thought to have a borderline personality disorder. At this facility, a staff person from each shift is required to make a chart entry; the abbreviations *DTC*, *ETC*, and *NTC* refer to day, evening, and night treatment counselors. Any other staff member who has contact with the client during the day (such as the therapist, teacher, nurse, or recreation specialist) also writes a staff note. The word *Level* refers to the number of privileges that a client (ct.) has. For example, a client at Level 1 has a bedtime of 9 p.m.; a client at Level 3 would go to bed at 10.

1/22/01, 8:15 A.M. DTC: Ct. awoke on time this morning. She was showered, dressed, and ready for school on time. She had positive interactions with her peers this morning. Ct. received one cue for being loud. Ct. completed her chores in a timely manner and responded appropriately to all staff requests. Ct. remains on Level 2 and under constant supervision. _____Jan Allen

---

1/22/01, 2:45 P.M. TEACHER: Ct. completed all of her classroom assignments in a timely manner. She received one C for cursing. She accepted the C appropriately and maintained a good attitude. Ct. had positive interaction with peers throughout the day. She had a slight confrontation with one peer, but the two of them worked it out in a positive manner. Ct. has been very talkative and cheerful today. Ct. responded appropriately to all staff requests. Ct. remains on Level 2 and under constant supervision. _____ Carlos Chaney

---

1/22/01 4:30 P.M. THERAPIST: Ct. requested and received her Level 3 today. She appeared very excited and stated that she has worked very hard for this and feels that she deserves it. We discussed the fact that being on Level 3 requires displaying Level 3 behavior. Ct. was receptive to this and seems to want to make the effort to do so. We also mutually decided that her new goal will be to take responsibility for her own actions and not blame other people. She understands that this is essential to her treatment and was receptive to doing so. _____ Kim Stuck

---

1/22/01 9:30 P.M. ETC: Ct. participated appropriately in all group activities this evening. Went outside for structured activity time and participated appropriately. Interacted positively with her peers. Ct. participated actively and appropriately in group. Discussed her new goal and how much she desires to achieve it. Gave positive and appropriate feedback in group. Ct. completed her chore in a timely manner without being prompted by staff. She discussed with staff how much she wants to go home and how she wants to work through the program as quickly as possible so that she can go home. She seemed a little quieter than usual this evening. Ct. required no cues for the evening and responded appropriately to all staff requests. Ct. remains on Level 3 and under constant supervision. _____ B. Greer

---

1/23/01 5:00 A.M. NTC: Ct. slept through the night with no problems and was present for each 15-minute bed check. _____ Phil Thress

---

This residential treatment facility gives its staff members strict guidelines for charting and consistent abbreviations: *ct* to mean client, *cue* to mean a warning, and *C* to signify a consequence. Also, there must be no blank lines or spaces in a case note, so the recorder puts in the line and signature at the end of each entry. Note also that no names of other clients or staff members appear in any entry. Notes such as these are routinely reviewed by the case manager.

You will also note that the word *appropriate* appears often. In this facility, it is an important word because these clients often behave and interact inappropriately. It is common for them to act out sexually, exhibit interpersonal difficulties, and rebel against any rules or authority figures.

 Chapter Summary

The assessment phase of case management includes the initial contact with the individual or individuals who need or desire services, the intake interview to gather data, and the documentation that is required. Evaluating the application for services concludes the phase.

Applicants learn about services in different ways, which often result in a referral. The first step in determining eligibility for services is the intake interview, which may be structured or unstructured. The intake interview is an opportunity to establish rapport, explore the need or problem, and give some structure to the relationship.

Case assignment follows acceptance for services and may occur in one of three ways. First, the case manager who conducts the intake interview may be assigned the case. Second, the case may be assigned to a specialized worker who may provide either complex services (level specialization) or specific services (task specialization). Finally, a team of professionals may be responsible for a number of clients.

Documentation and report writing are important parts of the assessment phase of case management. Two different types of documentation are process and summary recording. Intake summaries and case notes are examples of the documentation that occurs in this phase.

Chapter Review

◆ *Key Terms* ....

Applicant	*Process recording*
Client	*Summary recording*
Interview	*Confidentiality*
Structured interview	*Privileged communication*
Intake interview	*Intake summary*
Mental status examination	*Staff notes*
Unstructured interview	

◆ Reviewing the Chapter ....

1. How does a person's previous experience with human service agencies influence his or her decision to seek help?
2. In what different ways do people learn about available services?
3. Distinguish between the terms *applicant* and *client*.
4. What purposes does the interview serve?
5. Describe the roles of the helper and the applicant during the initial meeting.
6. Discuss different ways to define *interview*.
7. What is the difference between interviewing and counseling?
8. List the three factors common to all interviews.
9. Describe what takes place in each of the three parts of an interview.
10. Give three examples of issues that applicants may raise during the initial interview.
11. Why are intake interviews and mental status examinations structured interviews?
12. What does a helper need to know about confidentiality in the assessment phase?
13. Discuss the two problem areas related to confidentiality.
14. What questions guide the assessment of the information gathered during this first phase of service coordination?
15. Describe case assignment.
16. Compare process recording and summary recording, and provide an example of each.
17. What purposes do staff notes (case notes) serve?

◆ Questions for Discussion ....

1. Why do you think the interview is an important part of the case management process?
2. What evidence can you give that a good assessment is vital to the case management process?
3. Speculate on what could happen if a client's confidentiality were violated.
4. Develop a plan to interview a client who has applied for public housing.

References

Austin, M. J. (1981). *Supervisory management for the human services.* Englewood Cliffs, NJ: Prentice Hall.

Brill, N. I. (1998). *Working with people: The helping process.* White Plains, NY: Longman.

Corey, G., Corey, M. S., & Callanan, P. (1998). *Issues and ethics in the helping professions* (5th ed.). Pacific Grove, CA: Brooks/Cole.

Enelow, A., & Wexler, M. (1966). *Psychiatry in the practice of medicine.* New York: Oxford Press.

Epstein, L. (1985). *Talking and listening: A guide to the helping interview.* St. Louis: Times Mirror/Mosby.

Garrett, A. (1972). *Interviewing: Its principles and methods.* New York: Family Service Association of America.

Hutchins, D. E., & Cole Vaught, C. (1997). *Helping relationships and strategies* (3rd ed.). Pacific Grove, CA: Brooks/Cole.

Ivey, A. E., & Ivey, M. B. (1999). *Intentional interviewing and counseling: Facilitating client development in a multicultural society* (4th ed.). Pacific Grove, CA: Brooks/Cole.

McGowan, J. F., & Porter, T. L. (1967). *An introduction to the vocational rehabilitation process.* Washington, DC: U.S. Department of Health, Education, and Welfare.

McQuaide, S., & Ehrenreich, J. H. (1997). *Assessing client strengths. Families in Society,* 78(2), 201–212.

Okun, B. F. (2002). *Effective helping: Interviewing and counseling techniques* (5th ed.). Pacific Grove, CA: Brooks/Cole.

Perlman, H. H. (1979). *Relationship: The heart of helping people.* Chicago: University of Chicago Press.

Weinrach, S. G. (1987). The preparation and use of guidelines for the education of clients about the therapeutic process. *Psychotherapy in Private Practice,* 5(4), 71–83.

Wilson, S. J. (1980). *Recording: Guidelines for social workers.* New York: Free Press.

Chapter Five

Effective Intake Interviewing Skills

We do intake every Monday morning at 8:00. Sometimes clients are referred to us by the court system. They may parolees or on probation or waiting to be released. I do a pre-intake interview while they are still in custody. They come in handcuffs and shackles with two armed guards. I am looking for attitude, the presence of any communicable diseases, and a history of violence. I screen them very carefully because they may endanger the rest of the population here.
> —Oscar Ramirez, Miami Rescue Mission, personal communication, Miami, FL, May 13, 1997

With your first contacts with the family, you need to be able to ask the right questions. We have families who are very articulate and very forthcoming about what they are struggling with and what they need. And then we have many families who are not. . . . They don't know exactly what they need, or they can't articulate it for whatever reason. Our job then is to examine the whole situation, to gather every bit of information possible.
> —Margaret Mikol, SKIP of NY, personal communication, May 4, 1994

Listening is the most important tool. Not so much what you can tell them, but what they can tell themselves. Often a client will start talking to you and they will ask their questions and answer them in one breath. You have nothing to say because they will resolve their own problems. All you have to do is to be there and listen to them.
> —Sandra Flecha, caseworker, Casita Maria Settlement House, Bronx, NY, personal communication, May 4, 1994

Interviewing is described in the previous chapter as directed conversation or professional conversation. Many helpers consider it an art as well as a skilled technique that can be improved with practice. In case management, the intake interview is a starting point for providing help. Its main purpose is to obtain an understanding of the problem, the situation, and the applicant. A clear statement of the intentions of the interview helps the case manager and the client reach the intended outcomes.

The chapter-opening quotations illustrate some important skills that are needed during the interviewing process. At the Miami Rescue Mission, the interviewer screens by seeking specific information about attitude, communicable diseases, and violent behavior. Margaret Mikol begins her contact with families by using questions to grasp the big picture. She makes a distinction between questions in general and the *right* questions. Sandra Flecha, a caseworker at Casita Maria Settlement House, emphasizes listening as a critical skill in the interviewing process. It is her belief that clients will tell their problems if given an opportunity. Each of these professionals describes interviewing in a different way, but a common thread is respect for the considerable skills involved in using the interview to gain an understanding of the client's situation.

A number of factors influence interviewing in the helping professions. Some factors apply directly to the interviewer, such as attitudes, characteristics, and communication skills. Others are determined by the agency under whose auspices the interview occurs: the setting, the purpose of the agency, the kinds of information to be gathered, and recordkeeping. This chapter explores many of these factors.

The intake interview is usually the first face-to-face contact between the helper and the applicant. In some agencies, the person who does the intake interview will be the case manager; others have staff members whose primary responsibility is intake interviews. Interviews are also a part of the subsequent case management process, and some of the skills used in the intake interview apply there, too. This chapter uses the term *case manager* to refer to the helping professional who is conducting the interview.

This chapter is about effective interviewing in case management: the attitudes and characteristics of interviewers, the skills that make them effective interviewers, how these skills are used in structured interviews, and the pitfalls to avoid when interviewing. For each section of the chapter, you should be able to accomplish the following objectives.

ATTITUDES AND CHARACTERISTICS OF INTERVIEWERS
- List two reasons why the attitudes and characteristics of the case manager are important to the interview process.
- Name five characteristics that make a good interview.
- Describe a physical space that encourages positive interactions between the client and the case manager.
- List barriers that discourage a positive interview experience.

ESSENTIAL COMMUNICATION SKILLS
- List the essential communication skills that contribute to effective interviewing.
- List three interviewing skills.
- Support the importance of listening as an important interviewing skill.
- Offer a rationale for questioning as an art.
- Write a dialogue illustrating responses that a case manager might use in an intake interview.

INTERVIEWING PITFALLS
- Name four interviewing pitfalls.

Attitudes and Characteristics of Interviewers

The case manager's attitudes and characteristics as an interviewer are particularly important during the initial interview because this meeting marks the beginning of the helping relationship. Research supports the view that the personal characteristics of interviewers can strongly influence the success or failure of helping (Brown & Srebalus, 1996). In fact, Brammer (1993) concluded after a review of numerous studies that these personal characteristics are as significant in helping as the methods that are used.

One approach to the attitudes and characteristics of interviewers is a framework based on the work of Brammer (1993) and Combs (1969). It suggests that two sets of attitudes are important: one related to self and the other related to how one treats another person. Those related to self include self-awareness and personal congruence, whereas respect, empathy, and cultural sensitivity are among the attitudes related to treatment of another person. Elsewhere in the literature, other perspectives on helping attitudes and characteristics have as common themes the ability to communicate, self-awareness, empathy, responsibility, and commitment (Woodside & McClam, 2002).

The case manager communicates helping attitudes to the applicant in several ways, including greeting, eye contact, facial expressions, and friendly responses. The applicant's perceptions of the case manager's feelings are also important in his or her impression of the quality of the interview (Morales & Sheafor, 2001). Communicating warmth, acceptance, and genuineness promotes a climate that facilitates the exchange of information, which is the primary purpose of the initial interview. The following dialogue illustrates these qualities.

INTERVIEWER: *(stands as applicant enters)* Hello, Mr. Johnson *(shakes hands and smiles)*. My name is Clyde Dunn—call me Clyde. I'll be talking with you this morning. Please have a seat. Did you have any trouble finding the office?

APPLICANT: No, I didn't. My doctor is in the building next door, so I knew the general location.

INTERVIEWER: Good. Sometimes this complex is confusing because the buildings all look alike. Have you actually been to the Hard Rock Cafe in Cancun *(pointing to the applicant's shirt)*?

APPLICANT: No, I haven't. A friend brought me this T-shirt. I really like it.

INTERVIEWER: They certainly are popular. I see them all over the place. Well, I'm glad you could come in this morning. What seems to be the problem?

The case manager communicates respect for the applicant by standing and shaking hands. It is also easy to imagine that Clyde Dunn is smiling and making eye contact with Mr. Johnson. Clyde takes control of the interview by introducing himself, suggesting how Mr. Johnson might address him, and asking him to have a seat. His concern about Mr. Johnson finding the office and his interest in the T-shirt communicate warmth and interest in him as a person. Clyde also re-inforces Mr. Johnson's request for help in a supportive way. All these behaviors reflect an attitude on Clyde's part that increases Mr. Johnson's comfort level and facilitates the exchange of information.

The positive climate created by such a beginning should be matched by a physical setting that ensures confidentiality, eliminates physical barriers, and promotes dialogue. It is disconcerting to the applicant to overhear conversations from other offices or to be interrupted by phone calls or office disruptions. He or she is sharing a problem, and such events may lead to worries about the confidentiality of the exchange. Physical barriers between the client and the case manager (most commonly, desks or tables) also contribute to a climate that can interfere with relationship building. As much as the physical layout of the agency allows, the case manager should meet applicants in a setting where communication is confidential and disruptions are minimal. It is preferable to have a furniture arrangement that places the case manager and the applicant at right angles to one another without tables or desks between them and that facilitates eye contact, positive body language, and equality of position.

A sensitive case manager is also cognizant of other kinds of barriers, such as **sexism, racism, ethnocentrism**, and **ageism**. Problems inevitably arise if the case manager allows any biases or stereotypes to contaminate the helping interaction. To help you think about your own biases and stereotypes, indicate whether you believe each of the following statements is true or false.

T	F	Boys are smarter than girls when it comes to subjects like math and science.
T	F	Men do not want to work for female bosses.
T	F	Mothers should stay home until their young children are in school.
T	F	Women cannot handle the pressures of the business world.
T	F	Asians are smarter than other ethnic groups.
T	F	People on welfare do not want to work.
T	F	People who do not attend church have no moral principles.
T	F	A mandatory retirement age of 65 is necessary because people at that age have diminished mental capacity.
T	F	The older people get, the lower their sexual interest and ability.

How did you respond to these statements? Each statement reflects an unjustified opinion that is based solely on a stereotype of gender, race, or age.

Sensitivity to issues of ethnicity, race, gender, and age is important for the case manager when conducting interviews. Many clients and families have backgrounds very different from that of the case manager. In the United States today, one-fourth of the population originates from non-European backgrounds, a large number of clients are women, and the population proportion of elderly people is increasing rapidly. For many people in these populations, life is difficult, and they have few places to turn for help. Many of them live in poverty, have inadequate education, have a disproportionate chance of getting involved in the criminal justice system (either as a victim or a perpetrator), possess few useful job skills, are unemployed, and suffer major health problems at a disproportionate rate (Smith, 1995a).

Case managers should ask themselves, "How do I become sensitive to my clients and relate to them in a way that respects and supports their race, culture, gender, and age?" The following points may be helpful.

Each client is unique.   It is easy to stereotype cultural, racial, gender, or age groups, but clients cannot be understood strictly in terms of their particular culture. For example, many African Americans do share values and experience similar life events, but not all African Americans are the same. During interviews, case managers must take special care to get to know each individual client rather than categorizing him or her as a member of one particular group. For example, Marta Rivera, a caseworker at Casita Maria Settlement House in New York City, explained the complexities their workers face when working with Latino clients. "With Hispanics, it is just no longer the Puerto Rican family coming, but people from Nicaragua, Colombia, and Honduras come for help. All we share is that we are labeled Hispanics or Latino. . . . The dialects, the beliefs, the family lifestyles are totally different" (personal communication, May 4, 1994).

Language has different meanings.   Do not assume that words mean the same to everyone who is interviewed. When the case manager asks interview questions, clients sometimes do not understand the terminology. Likewise, words or expressions that clients use may have a very different meaning for the interviewer. For example, questions about family and spouse are familiar subjects in an intake interview. When clients talk about "partners" or "family," these terms can have various meanings, depending on the cultural background and life experiences of the individual being interviewed. For example, in the Native American culture, the *family* is an extended one that includes many members of the clan. For gay men and lesbian women, the word *partner* has the special meaning of "significant other."

Another example of language having different meanings is when working with a client who is deaf. One general rule of thumb is to avoid idioms and figurative language, such as "Cat got your tongue?" Someone who is hearing impaired may respond "Where is the cat?" after interpreting the phrase literally. A second general rule is to be aware of words with multiple meanings. For example, *hard* may mean difficult or it could mean rigid or unyielding. Words with multiple meanings are difficult for individuals with hearing impairment.

Explain the purpose of the intake interview and the case manager's role.   Clients may show up for the interview without understanding its purpose or the role of the interviewer in the helping process. Confidentiality may also be an important issue for them—sharing information about themselves and others may be contrary to the rules of their culture. For example, for many people raised in Asian cultures, to describe a problem to someone who is not in the family implies making the matter public, which is considered to bring shame to the family.

Clients may be different from you.   It is easy to make the mistake of expecting the clients we serve to be like us. We begin the interview process wanting to find similarities as a way of building a bridge to them. When clients prove to be very different, or we cannot understand them, we often want them to change so that they will be easier to "manage." In the United States, we often like to think of our country as a melting pot in which all cultures mix together and lose their original identities. When individuals do not want to lose their own culture, there is a tendency to blame them for being difficult. Case managers must take special care in the interview process to let clients know that there is respect for differences.

We have assembled some suggestions for developing sensitivity in interviewing individuals with certain cultural backgrounds (Baruth & Manning, 1999; Bernal et al., 1983; Gilligan, 1982; Kirst-Ashman, 1999; Smith, 1995a, 1995b). These are meant to be guidelines and points of awareness; they should be used with caution. As we mentioned earlier, individuals seldom exhibit all the characteristics of their cultural group.

Interviewing clients of Native American origin.   In many Native American cultures, sharing information about oneself and one's family is difficult. It is important not to give others information that would embarrass the family or imply wrongdoing by a family member.

Listening behaviors such as maintaining eye contact and leaning forward are considered inappropriate and intrusive in some Native American cultures.

For many Native Americans, trust increases as you become more involved in their lives and show more interest in them. Making home visits and getting to know the family can significantly improve an interviewer's chances of getting relevant information.

Native Americans tend not to make decisions quickly. The slowness of the process could influence how soon the client is willing to share information or make judgments.

Native American cultures sometimes incorporate a fatalistic element—a belief that events are predetermined. During the initial stages of the process, the client may not understand how his or her responses and actions can influence the course of service delivery.

Interviewing clients with a common background of Spanish language and customs.   Individuals living in the United States who are of Mexican, Central and South American, or Caribbean ancestry are often referred to as Hispanic, Latino, and Chicano. There is actually little agreement on the appropriate term for identification across groups and even within subgroups. Although they share some commonalities (to be discussed here), they may differ in appearance, country of origin, date of immigration, location and length of time in the United

States, customs, and proficiency in English. Case managers should be sensitive to terminology and avoid stereotypes.

Many cultures with this common background view informality as an important part of any activity, even the sharing of information. Taking time to establish rapport with the client before direct questioning begins is helpful.

Some people of this origin may be perceived as submissive to authority because they appear reticent or reluctant to answer questions. Their behavior, in fact, may be shyness or the natural response to a language barrier.

The father may be seen as aloof as he performs his roles of earning a living for the family and establishing the rules. The mother and other members of the family tend to assume more nurturing roles. Questions that do not take these roles into consideration may be misinterpreted by the clients or may suggest to them that the interviewer is an outsider incapable of understanding the culture or of helping them.

Fatalism often plays a role in these cultures. These clients may not see any point in discussing the future, preferring to talk about the present.

Interviewing African Americans.   Given a long history of being denied access to services, African Americans may not want to share information during the intake interview. They may be angry about discriminatory treatment that they have experienced in the human service delivery system. During the intake interview, it is important to focus on concrete issues that can be connected to services. This approach shows respect for the client's right to expect fair treatment and quality services.

When being interviewed by a white professional, an African American may feel powerless because of previous experiences in white society. The interviewer will need to show that the client's input does matter.

Interviewing women.   Many women do not know how to talk about the difficulties that they are experiencing, and they may not know how to respond to the questions they are asked. Some have had few opportunities to discuss their problems and may believe they do not have the right to complain. Listening carefully is very important.

Anger may play a part in the initial interview. Many women come to the helping process frustrated, either because their efforts have been unrecognized or because they believe that others expect them to be perfect. Often this anger must be expressed before any information can be gathered.

Women often feel powerless and do not expect the bureaucracy to serve them well. They may be reluctant to communicate and doubtful that the interview or the process as a whole can make a difference.

Women may also fill different roles in their lives that may conflict or cause confusion. When interviewing about client strengths, women from some traditional cultures in the United States may defer to males and elders and subordinate their own individuality, yet at work and at school, they may be assertive and confident (Gil & Vasquez, 1996). Without exploration, these differences may be perceived as weaknesses while in fact, the flexibility and role shifts may be strengths. Learning about roles and demands contributes to an understanding of the client's situation.

Women may be overly dependent as clients and assume that the case manager will take complete control of the interview. They may want the interviewer to be the one to identify problems and possible goals. In such cases, care should be taken to give the woman opportunities and encouragement to respond more fully.

Interviewing elderly clients.   In this society, elderly people are often disregarded and devalued. During the interview, the case manager must show respect for the elderly client's answers and opinions about the issues discussed. Such a client needs to be assured that his or her responses are important and have been heard by the helper.

Pay special attention to the elderly client's description of support in his or her environment. Many live in an environment of decreasing support (changing neighborhood, death of friends) and with decreasing mobility. Others live with limited family support. These clients may not realize how their environment has changed.

Elderly clients may be reluctant to share their difficulties for fear of losing much of their independence. They may understate their needs or overstate the amount of support they have, hoping to avoid changes in their living conditions, such as being removed from their homes or relinquishing their driving privileges.

Interviewing individuals with disabilities.   Although individuals with disabilities are not traditionally considered a cultural group, it is important to develop a sensitivity to the issues these individuals may encounter. Attitudes toward people with disabilities are often based on the amount of information and education about disabilities and on the amount of contact a person has had with individuals with disabilities (Eichinger, Rizzo, & Sirotnik, 1992). These factors are also the best predictors of positive attitudes toward people with disabilities (Yuker, 1994). Case managers working with this population need to know about mental, physical, and emotional disabilities; the onset of disability; acceptance of disability; disabilities as handicaps; accommodations; and treatment.

A major source of information about a disability is the client. As with other clients, establish the helping relationship by building rapport and trust. Then address the disability or condition: Is it the problem? If not, does it affect the problem? Is it even related to the reason the client is seeking services? Don't make assumptions about why the individual is there or about the disability, and don't generalize. Each person is unique.

Case managers also need to increase self-awareness about their own attitudes and knowledge. Know your limits and control your reactions. Increase your knowledge by learning from your client about a particular disability, the difficulties faced, and the environmental situations that are problematic.

These are only a few of the differences that helpers may encounter during the intake interview with individuals of various ethnic, racial, gender, and age groups. In several ways, case managers can continue to learn more about how to interview culturally diverse clients. Among them are becoming knowledgeable about other cultures, reading professional articles that focus on ways to modify the interviewing process to meet the needs of certain client groups, and talking with other helpers whose own cultural origins give them insight into cultural barriers. Gaining an understanding of diversity is a process that continues throughout the professional life of every effective helper. Such an understanding enhances the interviewing environment for both parties.

Essential Communication Skills

Communication forms the core of the interviewing process. In interviewing, communication is the transmission of messages between applicant and helper. As the first face-to-face contact, the interview is a purposeful activity for both participants. In many cases, the motivation is a mutual desire to decide whether the applicant is in the right place for the needed services. This is a negotiation that is facilitated by **effective communication skills**.

An important skill that promotes the comfort level of the applicant and lays the foundation for a positive helping relationship is using language the person understands. This means avoiding the use of technical language. For example, terms such as *eligibility*, *resources*, and *Form 524* may not mean much to an applicant who has not become familiar with the human service system. To take another example, imagine that the interviewer is discussing the benefits of taking a vocational or interest test. Rather than going into detail about the validity or reliability of the test, the case manager should discuss how it might help in establishing a vocational objective. Using language or words the applicant does not understand tends to create distance and disengagement.

Congruence between verbal and nonverbal messages is another way to facilitate the interaction between an applicant and a case manager. A major part of the meaning of a message is communicated nonverbally, so when conflict is apparent between the verbal and nonverbal messages, the applicant is likely to believe the nonverbal message. A common example of this is the person who says, "Yes, I have time to talk with you now," while dialing the phone or looking through her desk drawer for a folder. The lack of eye contact or any other encouraging nonverbal message communicates to us that the person is indeed busy or preoccupied with other matters.

Another skill that facilitates the interview process is **active listening**—making a special effort to hear what is said, as well as what is not said. An interviewer who is sensitive to what the applicant is communicating, verbally as well as nonverbally, gains additional information about what is really going on with the individual. This ability is particularly helpful in situations in which the presenting problem may differ from the underlying problem and when interviewing an individual from another culture. Later in the chapter, we present a more detailed discussion of listening as it relates to the intake interview.

A popular way to elicit information is by asking questions. **Questioning** is an art as well as a skill. Unfortunately, case managers don't often develop their questioning skills, relying instead on questioning techniques that have served them well in informal or friendly encounters. Typically, this means asking questions that focus on facts, such as "What happened?" "Who said that?" "Where are you?" "Why did you react that way?" Questions such as these usually lead to other questions, placing the burden of the interview on the case manager and allowing the applicant to settle into a more passive role. The applicant's participation is then limited to answering questions, so the interview may begin to feel like the game "Twenty Questions." Skillful questioning combined with effective responding helps elicit information and keep the interaction flowing. Appropriate questioning and responding techniques are introduced later in this chapter.

Patterns of communication vary from culture to culture, according to religion, ethnic background, gender, and lifestyle differences. In the dominant culture in the United States, it is effective to use a reflective listening approach when feelings are important. Many of the techniques that are useful in this approach are not appropriate for all cultures. For example, eye contact is inappropriate among some Eskimos. The sense of space and privacy is different for Middle Easterners, who often stand closer to others than Americans do. Some people from Asian cultural backgrounds may prefer more indirect, subtle approaches of communication. Thus, a single interviewing approach may have different effects on people from various cultural backgrounds. The skillful and sensitive case manager must be aware of these differences.

Both spoken language and body language are expressions of culture. Many helpers work with clients from several cultures, each with their own assumptions and ways of structuring information. Both talking and listening provide many occasions for misunderstanding (Kolanad, 1994). Jan Rasmussen, at St. Patrick's in St. Louis, Missouri, talks about the necessity of understanding various cultural backgrounds. "Sixty-six percent of our clients are male and 34% percent are female. Seventy-five percent are African Americans, 17% White and 6% Hispanic . . . we have a few Asians and Native Americans" (personal communication, March 6, 2002). Their staff is continually learning about the meaning of words and expressions used by their diverse set of clients. When they do not understand or are lost in the communication, then they ask the clients to help them understand.

Assigning great significance to any single gesture by the applicant is also risky, but a pattern or a change from one behavior to another is meaningful (Sielski, 1979). Once again, the key is the case manager's awareness during the interview process.

Now that you have read about general guidelines for essential communication, let's focus on the specific skills of listening, questioning, and responding.

Interviewing Skills

Interviewing skills aim to enhance communication, which involves both words and nonverbal language. Spoken language varies among individuals and cultures. Understanding spoken language is challenging because it is always changing, it is usually not precise, and it is ambiguous. Body language, which is also important and challenging to understand, includes body movement, posture, facial expression, and tone of voice. Knowing the ways in which body language varies culturally can help the interviewer fathom the thoughts and feelings of the applicant.

In talking with an applicant, the case manager must strive for effective communication, making sure that the receiver of the message understands the message in the way the sender intended. In the intake interview, the case manager listens, interprets, and responds. To understand the applicant's problem as fully as possible, the case manager constantly interprets the meanings of behaviors and words. He or she should always have a "third ear" focused on this deeper interpretation.

At the same time, the applicant is interpreting the words and behaviors of the case manager. A case manager who is an effective interviewer can help the applicant make connections and interpretations. Also contributing to correct interpretations and connections is a good working relationship between the two of them, good timing, and sensitivity to whether the material being discussed is near the applicant's level of awareness.

Zulma Resto, a caseworker at Casita Maria Settlement House, describes the initial meeting at her agency.

> Normally when clients come in, they are already tired of the system, and so when they meet you, they are like ticked off and probably hate you. One thing that I try to get across right away is that I have nothing to do with the welfare system, which is what most of them are against. I let them know that I am here for them and that whatever we say is confidential. I try to keep the first session free of paperwork. Keep it friendly and not get too much into their space the first day. Then the second time they see me, they know who I am and we can talk a little bit more. It's important to bond with them, get their confidence and trust, so that they will tell you what their real problems are. (Personal communication, May 4, 1994)

Sgt. Richard Valdemar, a gang specialist with the Los Angeles Sheriff's Department, explains how he establishes a relationship with gang members.

> I almost never ask about crime right away. First, I talk to them about anything else—cars, girlfriends, their neighborhood, clothes—just to let them get used to talking to me to see if we have something in common. After that, they loosen up and start telling me what's really happening. At the end, I like to ask them what they would be if they hadn't become a gang member. You know what they say? "Policeman. I'd like to be a policeman or maybe a probation officer." (Personal communication, March 25, 1998)

Both of these helping professionals are experienced at intake interviewing. They value the helping relationship and recognize its importance in the service delivery that is to follow. To establish the relationship, they use communication skills, such as listening, questioning, and responding. These are discussed and illustrated next, with excerpts from intake interviews.

LISTENING

Listening is the way most information is acquired from applicants for services. The case manager listens to the applicant's verbal and nonverbal messages. "Listening with the eyes" means observing the client's facial expressions, posture, gestures, and other nonverbal behaviors, which may signal his or her mood, mental state, and degree of comfort. Verbal messages communicate the facts of the situation or the problem and sometimes the attendant feelings. Often, however, feelings are not expressed verbally, but nonverbal messages provide clues. A good listener should be sensitive to the congruence (or lack of it) between the client's verbal and nonverbal messages. The case manager must pay careful attention to both forms of communication.

Good listening is an art that requires time, patience, and energy. The case manager must put aside whatever is on his or her mind—whether that is what to recommend for the previous client, the tasks to be accomplished by the end of the day, or making a grocery list—to focus all attention on the applicant. The case manager must also be sensitive to the fact that his or her behavior gives the applicant feedback about what has been said (Epstein, 1985). During the interview, the case manager must also recognize cultural factors that play into the interpretation of body language. For example, the proper amount of eye contact and the appropriate space between case manager and applicant may vary according to the cultural identity of the applicant. As you can see, listening is indeed complicated. What behaviors characterize good listening? How are attentiveness and interest best communicated to the applicant?

Attending behavior, *responsive listening*, and *active listening* are terms that indicate ways in which case managers let applicants know that they are being heard. The following five behaviors are a set of guidelines for the interviewer (Egan, 1998, pp. 63–64). They can be easily remembered by the acronym S-O-L-E-R.

S: Face the client *squarely*; that is, adopt a posture that indicates involvement.

O: Adopt an *open posture*. Crossed arms and crossed legs can be signs of lessened involvement with or availability to others. An open posture can be a sign that you're open to the client and to what he or she has to say.

L: Remember that it is possible at times to *lean* toward the other. The word *lean* can refer to a kind of bodily flexibility or responsiveness that enhances your communication with a client.

E: *Maintain good eye contact*. Maintaining good eye contact is a way of saying, "I'm with you; I'm interested; I want to hear what you have to say."

R: Try to be relatively *relaxed*. Being relaxed mean two things. First, it means not fidgeting nervously or engaging in distracting facial expressions. Second, it means becoming comfortable with using your body as a vehicle of personal contact and expression.

Attending behavior is another term for appropriate listening behaviors. Eye contact, attentive body language (such as leaning forward, facing the client, facilitative and encouraging gestures), and vocal qualities such as tone and rate of speech are ways for the interviewer to communicate interest and attention (Ivey & Ivey, 1999). Attending behavior also means allowing the applicant to determine the topic.

Other guidelines for good listening are provided by Epstein (1985, pp. 18–19).

1. Be attentive to general themes rather than details.
2. Be guided in listening by the purpose of the interview in order to screen out irrelevancies.
3. Be alert to catch what is said.
4. Normally, don't interrupt, except to change the subject intentionally, to stop excessive repetition, or to stop clients from causing themselves undue distress.
5. Let the silences be, and listen to them. The client may be finished, or thinking, or waiting for the practitioner, or feeling resentful. Resume talking when you have made a judgment about what the silence means, or ask the client if you do not understand.

A skillful listener also hears other things that may help him or her understand what is going on. A shift in the conversation may be a clue that the applicant finds the topic too painful or too revealing, or it might indicate that there is an underlying connection between the two topics. Another consideration is what the applicant says first. "I'm not sure why I'm here" or "My probation officer told me to come see you" give clues about the applicant's feelings about the meeting. Also, the way the applicant states the problem may indicate how he or she perceives it. For example, an applicant who states, "My mother says I'm always in trouble," may be signaling a perception of the situation that differs from the mother's. Concluding remarks may also reveal what the applicant thinks has been important in the interview. The skilled interviewer also listens for recurring themes, what is not said, contradictions, and incongruencies.

Good listeners make good interviewers, but as you have just read, listening is a complex activity. It requires awareness of one's own nonverbal behaviors, sensitivity to cultural factors, and attention to various nuances of the interaction. It is further complicated by the fact that people seeking assistance don't always say what they mean or behave rationally. However, the use of good listening skills always increases the likelihood of a successful intake interview.

QUESTIONING

Questioning, a natural way of communicating, has particular significance for intake interviews. It is an important technique for eliciting information, which is a primary purpose of intake interviewing. Many of us view questioning as something most people do well, but it is in fact a complex art. This section elaborates on questioning skills, introduces the appropriate use of questions, identifies problems that should be considered, and explores the advantages of open inquiry as one way to elicit information.

Questioning is generally accepted by some as low-level or unacceptable interviewer behavior (Carkhuff, 1969; Egan, 1998; Gordon, 1974). Others view it as a complex skill with many advantages (Ivey & Ivey, 1999; Long, Paradise, & Long, 1981). Let's explore its complexity and its advantages. Long, Paradise, and Long (1981) give three reasons that questioning is a complex skill: Questioning may assist *and* inhibit the helping process; it can establish a desired as well as an undesired pattern of exchange; and it can place the client in the one-sided position of being interrogated or examined by the helper.

For these reasons, we may consider questioning an art form. The wording of a question is often less important than the manner and tone of voice used to ask it. Suzy Bourque at the Family Counseling Agency in Tucson says: "I think that people have to be detectives. . . . They have to enjoy walking into a new setting and seeing what is there. . . . It is not just going out with your 12-page assessment form and asking alienating questions" (personal communication, October 7, 1994). Linda Smith at the School for the Deaf concurs: The "skills are those involved with being a private eye, nosy in a tactful way" (personal communication, October 27, 1993).

Also, too many questions will confuse the applicant or produce defensiveness, whereas too few questions place the burden of the interview on the client, which may lead to the omission of some important areas for exploration. The pace of questions influences the interview, too. If the pace is too slow, the applicant may interpret this as lack of interest, but a pace that is too fast may cause important points to be missed. A delicate balance is required.

What are the advantages of questioning? One is that questioning saves time. If the case manager knows what information is needed, then questioning is a direct way to get it. Questioning also focuses attention in a particular direction, moves the dialogue from the specific to the general as well as from the general to the specific, and clarifies any inaccuracies, confusion, or inconsistencies. Let's examine some examples of the appropriate uses of questioning. After each example, you are asked to provide two relevant questions.

To begin the interview Could you tell me a little about yourself? What would you like to talk about? Could we talk about how I can help?

You work at the county Office on Aging. A woman comes in with her elderly mother. List two questions that you might use to begin the interview.

_____

_____

_____

_____

To elicit specific information How long did you stay with your grandmother? What happens when you refuse to do as your boss asks? Who do you think is pressuring you to do that? Can you give me an example of a time when you felt that way?

A client tells you about mistreatment by her boss at her new job. She claims that she is being sexually harassed. What two questions would you ask to help you understand what happened?

To focus the client's attention Why don't we focus on your relationship with your daughter? What happens when you do try to talk to your husband? Of the three problems you've mentioned today, which one should we discuss first?

A client is worried about how her surgery will go, who will care for her children while she is in the hospital, and whether she will be fired for missing so much work. She wrings her hands and seems ready to burst into tears. What are two questions you could use to focus her attention?

To clarify Could you describe again what happened when she left? How did you feel about that conversation compared with others you have had with him? What is different about these two situations?

A young man shares his anguish over his mother's death a year ago. You notice that he is smiling, and you are confused about what he is really saying. Write two questions that would help you clarify what is going on.

To identify client strengths What is a current problem you have also faced in the past? Can you now use the same resources to solve your current problem? What did you do to keep the problem from turning into a crisis?

A family member with a disability is questioned to assess functioning level and suitability for a program that requires her to ride public transportation. Write two questions that would help you identify her strengths.

These are examples of interview situations in which the case manager might legitimately use questions. In all of them, the general rule of questioning applies: Question to obtain information or to direct the exchange into a more fruitful channel.

Although questioning may seem to be the direct path to information, sometimes this strategy can have negative effects. Long, Paradise, and Long (1981) suggest that interviewers not rely on questions to carry the interaction or interview. This is particularly problematic for beginning helpers because people generally have a tendency to ask a question whenever there is silence (discussed later in this section). Questions may also be inappropriate when the case manager does not know what to say. Asking questions nervously may lead to more questions, which can put the case manager in the position of focusing on thinking up more questions rather than listening to what the client is saying. Prematurely questioning to assess client strengths during the interview can also be problematic and may be viewed as rejection by the client.

An overreliance on questioning can create other problems for both interviewers and clients. For the client, too many questions can limit self-exploration, placing him or her in a dependent role in which the only responsibility is to respond to the questions. A client may also begin to feel defensive, hostile, or resentful at being interrogated. Using too many questions may place the case manager in the role of problem solver, giving him or her most of the responsibility for generating alternatives and making decisions. In the long term, overreliance on questioning leads to bad habits and poor helping skills. Using questions to the exclusion of other types of helping responses eventually results in the withering of these other skills (as discussed in the next section).

In conclusion, questioning is an important strategy for effective interviewing, but it is more than a strategy for obtaining information. Because of the subtleties of questioning, the matter of its appropriate uses in interviewing, and the potential problems, questioning is an art that requires practice. The skillful case manager who uses questioning to best advantage knows when to use open and closed inquiries to gather information during the intake interview. These types of questions are discussed next.

CLOSED AND OPEN INQUIRIES

The questions used in intake interviews can be categorized as either open or closed inquiries. Determining which one to use depends on the case manager's intent. If specific information is desired, closed questions are appropriate: "How old are you?" "What grade did you complete in school?" "Are you married?" If the case manager wants the client to talk about a particular topic or elaborate on a subject that has been introduced, open questions are preferred: "What is it like being the oldest of five children?" "Could you tell me about your experiences in school?" "How would you describe your marriage?"

Closed questions elicit facts. The answer might be yes, no, or a simple factual statement. An interview that focuses on completing a form generally consists of closed questions like those in the previous paragraph. However, the interviewer must be cautious, for a series of closed questions may cause the client to feel defensive, sensing an interrogation rather than an offer of help. One approach is to save the form until the end of the interview, review it, and complete the unanswered questions at that time. If the completion of an intake form is allowed to take precedence in the interview, the case manager misses the opportunity to influence the client's attitudes toward the agency, getting help, and later service provision. Perhaps equally important, information that could be acquired through listening and nonverbal messages may be missed if the interviewer is focused on writing answers on the intake form.

Open inquiries, on the other hand, are broader, allowing the expression of thoughts, feelings, and ideas. This type of inquiry requires a more extensive response than a simple yes or no. The exchange of this type of information contributes to building rapport and explaining a situation or a problem. Consider the following example.

FATHER:	I'm having trouble with the oldest boy, William. He's in trouble again at school.
INTERVIEWER 1:	How old is William?
INTERVIEWER 2:	Could you tell me more about what's going on?

Interviewer 1's response is a closed question that asks for a simple factual answer. Interviewer 2's response is an open inquiry that asks the father to elaborate on what he thinks is happening with William. This allows William's father to determine what he wishes to tell the interviewer about the situation. Such an open inquiry emphasizes the importance of listening—to what the individual says first, how he or she perceives the problem, and what is considered important.

You can see how valuable open inquires can be in intake interviewing. They also provide an opportunity for the clients to introduce topics, thereby putting them at ease by allowing discussion of their problems in their own way and time. Besides providing the information that the case manager needs, open inquiries encourage the exploration and clarification of the client's concerns.

Four methods are commonly used to introduce an open inquiry (Evans, Hearn, Uhlemann, & Ivey, 1998). Each is presented here with an example of a client statement, the interviewer's response, and the kind of information that the client might volunteer in response to the open inquiry.

- "What" questions are fact oriented, eliciting factual data.

MR. CAGLE:	I'm here to get food stamps. Here's my application.
INTERVIEWER:	Let's review it to make sure you've completed it correctly. What's your income?
MR. CAGLE:	Well, I make minimum wage at my job, and my wife don't make much either. We have three children and we live in a low-income apartment.

- "How" inquiries are people oriented, encouraging responses that give a personal or subjective view of a situation.

TAMISHA:	My boyfriend doesn't like my parents, and when we are all together, nobody agrees with anyone about anything.
INTERVIEWER:	How do you feel about that?
TAMISHA:	I hate it. Everyone is so uncomfortable. I want everyone to get along, but I dread the times we have to be together. Sometimes I feel like somebody will yell at someone else or even hit somebody.

- "Could," "could you," or "can you" are the kinds of open inquiries that offer the client the greatest flexibility in responding. These inquiries ask for more detailed responses than the others.

JUAN:	I hate school. My teacher doesn't like me. She's always on my case about stuff.
INTERVIEWER:	Could you describe a time when she was on your case?
JUAN:	Well, I guess. Like yesterday, she was mad at me because I was late to class . . . but I was only five minutes late. Then she called on me to answer a question. Well, I hadn't read the stuff because I lost the book, so how could I answer the question? I mean, give me a break.

The fourth type of open inquiry is the "why" question, which experienced interviewers often avoid because it may cause defensiveness in clients. Examples of "why" questions that may do this are "Why did you do that?" and "Why did you think that?" Phrased this way, these responses may be perceived as judgments that the client should not have done something, felt a certain way, or had certain thoughts. Less risky "why" questions are those phrased less intrusively: "Why don't we continue our discussion next week?" "Why don't we brainstorm ways that you could handle that?"

In what follows, we analyze some excerpts from an intake interview that occurred at juvenile court. Tom Rozanski is the case manager who was assigned to court on that particular day. In some such cases, the juvenile is remanded to state custody that very day, that is, he or she can leave the courthouse only to go to a local or state facility. The juvenile in this case, Jonathan Douglas, has been charged with breaking and entering. He has a history of substance abuse and school truancy and is well known to the judge, who finds him guilty and remands him to state custody. The case then comes under the jurisdiction of an Assessment, Care, and Coordination Team (ACCT), which takes responsibility for assessing the case, developing a plan of services, and coordinating the needed services among the agencies that are involved with the plan. Tom finds on this day that court is very crowded. Once Jonathan Douglas has been remanded to state custody, Tom asks him to follow him into the hall, and the initial intake interview occurs there. Jonathan's parents also join them, as do two officers, who suspect that Jonathan will run if he gets the chance. They stand together in the hall for a brief interview so that Tom can gather enough information to arrange a placement that afternoon. Here's what happens.

TOM: Jonathan, my name is Tom. (*Shakes hands*) I work for the As-
                   sessment, Care, and Coordination Team. We are responsible
                   for assessing your case and planning services for you.

JONATHAN: (*Limply shakes hands and looks everywhere but at Tom*)

TOM: Jonathan, are you listening? Please look at me. Are you on any
                   drugs right now?

JONATHAN: (*Unintelligible response*)

Tom realizes that it is futile to try to talk with Jonathan now and hopes that in a
few hours he will be down from whatever drugs he has taken.

TOM: Mr. and Mrs. Douglas, I am Tom Rozanski, a case manager
                   for the Assessment, Care, and Coordination Team in this
                   county. Let me review for you what has happened. The
                   judge found Jonathan guilty of breaking and entering. Be-
                   cause of his prior record, he is in state custody, and it is
                   my job to find a place for him to stay while we evaluate
                   his case. I need some basic information right now. Can
                   you help me?

MRS. DOUGLAS: Yes, we want to help him any way we can.

TOM: Does Jonathan live with either of you?

MRS. DOUGLAS: He stays with me once in a while, but mostly he stays
                   with his dad.

TOM: Mr. Douglas, could you describe his behavior when he
                   stays with you?

MR. DOUGLAS: Well, I guess he goes to school sometimes. Leastways,
                   when I leave for work, I try to get him up. I don't know if
                   he goes, though. Sometimes he's here when I get home
                   and sometimes he isn't. He's a big boy now, and I can't do
                   much with him, so I just let him be.

TOM: Do either of you have health insurance?

BOTH PARENTS: No.

The interview lasts approximately 5 more minutes, and Tom obtains some key information about the family situation. He has very little time and needs specific information, so he hurriedly asks closed questions. "What is your address, Mr.
Douglas?" "What grade is Jonathan in?" "Has he had a medical examination recently?" Finally, Tom has enough information to complete most of the intake
form. That afternoon, he meets with Jonathan and makes another attempt to talk
with him. He is relieved to find Jonathan more communicative at this meeting.
Here's an excerpt; note Tom's use of open inquiries.

TOM: Jonathan, I would like to talk with you about what's going
                   to happen. I'd also like you to tell me your side of what's
                   going on.

JONATHAN: (*Looks at Tom but makes no comment*)

TOM:	When we finish talking, Deputy Johnston will take you to Mountainview Hospital, where you will spend the next two weeks. During that time, we will talk again, you will take some tests, and you will meet with a group of young people who are your age. At the end of that time, we will develop a plan of services for you. Now, could you tell me about yourself?
JONATHAN:	Well, I'm 15. I don't like school and I don't get along with either of my parents. My mother doesn't want me since she moved, and my dad don't care if I'm at home or not.
TOM:	This is the first time you have been in trouble for breaking and entering. What happened?
JONATHAN:	Well, I was with these guys and we needed money for some dope. It looked easy. I think I made a mistake.
TOM:	Yeah. It seems so. Let's talk about what you can do now. What kind of changes would you like to see?
JONATHAN:	Well, I don't want to go to jail and I don't want to go to Red River [a juvenile correctional facility]. I can't stay home though. They don't care about me and I don't care about them.
TOM:	How would you describe your relationship with your parents?
JONATHAN:	We don't have no relationship. They don't care about me. Sometimes I stay with my mom, but she's looking for another husband and she don't want me around. My dad, he just don't want to be bothered.
TOM:	Hmm. Sounds as though you're not sure if there's a place for you with them. What changes would you like to see in your relationship with your parents?
JONATHAN:	I wish they . . . I wish . . . I wish they liked me.
TOM:	I see. Could you give me an example of what they would do if they liked you?
JONATHAN:	I don't know.
TOM:	Can you describe a time when you did something they liked?
JONATHAN:	My mom likes it when I come in early. My dad, he don't care.
TOM:	What have you done to please your mom?
JONATHAN:	(Pauses) I cleaned up the kitchen once.

In this excerpt, Jonathan mentions his family and school in his first response. Tom picks up on the family situation and decides to explore it with Jonathan. He has talked with the parents, and although they are not living together, he senses that both are interested in Jonathan and willing to help him but don't seem to know what to do, and they feel that Jonathan rebuffs any overtures they make. Tom is trying to discover what kind of support may be available to Jonathan from his parents and how receptive he would be to it. Tom uses open inquiries in his conversation with Jonathan to elicit the boy's thoughts and feelings about this issue. The use of "what" questions gets at factual information, and the "how" questions are aimed at people-oriented information.

In summary, case managers who are good interviewers use both open and closed inquiries, although open inquires are preferred whenever possible. They are also careful to ask one question at a time and to avoid asking consecutive questions of a kind that might create the feel of a cross examination. What other types of responses do interviewers use? The next section suggests other ways of responding to clients in an interview situation.

RESPONDING

A case manager might use various kinds of responses during the course of an intake interview. Of course, the type of response depends on the intent at that particular point. Let's review some of the most common responses. In the following material, each response is followed by an example of its use. Joe Barnes, a recent parolee, has returned home and is having a difficult time with his wife. His parole officer, sensing that the relationship is in trouble, suggests that Joe see a counselor at the Family Service Center.

Minimal responses   Sometimes called **verbal following**, minimal responses let the client know that you are listening. "Yes," "I see," "Hmm," and nodding are minimal responses. Using them is important when getting to know the applicant.

> JOE BARNES: I'm here because my probation officer thought it would be a good idea for me to talk with someone about things at home. Things haven't been very good since I came home.
>
> MIKE MATSON: I see.

Paraphrase   This response is a restatement (in different words) of the main idea of what the client has just said. It is often shorter and can be a summary of the client's statement. Paraphrasing lets the client know that the case manager has absorbed what was said.

> JOE: I just don't know what the trouble is. I was glad to get home, and I thought my wife would be glad to have me there. But we fight about everything—even stuff like when to feed the dog. I don't know what to do.
>
> MIKE: You don't know what's happening between you and your wife since you got home. Sounds like it's pretty unpleasant for both of you . . . and you're wondering what to do about it.

Reflection   Sometimes people get out of touch with their feelings, and reflection can help them become more aware. The feelings may not be named by the individual but, rather, communicated through facial expression or body language. For example, a flushed face or a clenched fist may show anger. The case manager's reflective response begins with an introductory phrase ("You believe," "I gather that," "It seems that you feel") and then clearly and concisely summarizes the feelings the case manager perceives.

> JOE: Yes. I don't know how we can continue to live like this. I know she is really angry about me getting in trouble with the law, but I've paid my dues, learned my lesson. I don't plan to ever get in that mess again.
>
> MIKE: I gather that you really do feel bad about what you did, but you would like to put the past behind you and focus on the future and how to make your marriage work.

Reflection is a response that facilitates a discussion of the client's feelings, particularly when he or she may feel threatened by such a discussion. It is also helpful as a way to check and clarify the case manager's perception of what was said during the interview.

Clarification   Clarifying helps the case manager find out what the client means. When the case manager is confused or unsure about what has taken place, it is more productive to stop and clarify at the time than to continue.

> JOE: I got so angry last week because she wouldn't listen to me and she didn't seem to care that I was home. I was yelling, she was yelling, she threw a bowl at me, and I almost hit her.
>
> MIKE: Sounds to me like you got so angry and frustrated that you were almost out of control.

Summarizing   With this response, the interviewer provides a concise, accurate, and timely summing up of the client's statements. It also helps organize the thoughts that have been expressed in the course of the interview. Summarizing is used to begin an interview when there is past material to review. It is also useful during the interview when a number of topics have been raised. Summarizing directs the client's attention to the topics and provides direction for the next part of the interview.

From the summary, the client can choose what to discuss next. Summarizing is also useful when the client presents a number of unrelated ideas or when his or her comments are lengthy, rambling, or confused; such a response can add direction and coherence to the interview. Finally, summarizing is a way to close the interview: The case manager goes over what has been discussed. Prioritizing next steps or topics becomes easier at this point.

> JOE: I told her I didn't care what she thought. I'm sure she knew what I meant even though I didn't know what I meant. She won't give me a chance. I am trying hard, so what does it mean to her that I have been gone? She has no idea what I have been through.
>
> MIKE: Let me see if I can summarize what we've talked about today. Returning home has been very difficult for you and you're confused about your relationship with your wife. She still seems angry about your trouble with the law, and the two of you just can't seem to communicate.

JOE: I guess that's about it.

MIKE: Let's focus on the communication problems at our next meeting.

The following is an excerpt from an intake interview that incorporates all the responses that you have just read about: open and closed inquiries, minimal responses, paraphrases, reflection, clarification, summarization, interpretation, confrontation, and informing. Notice how and when the case manager uses each response and the client's reaction to it.

Mathisa walked into the AIDS Community Center one Wednesday evening about 8 o'clock. She had come to talk to a counselor because she had just discovered that her best friend had AIDS. Her friend had told Mathisa and no one else, and Mathisa was scared. She did not know what to tell her friend, and she did not know what to do. Mathisa passes by the center on her way to school each morning, but she had barely noticed it. And now she was here.

A young man came up to her and introduced himself. She said "Hi" but did not want to tell him her name. In fact, she really did not want anyone to know that she was there. He asked her if she had come to talk and she nodded. He led her into a small room that had three comfortable chairs. He sat in one and pointed to one where she could sit.

The young man, Dean, started by telling Mathisa about the agency and about his job as a service coordinator. He also talked to her about the confidentiality policies of the agency.

DEAN: *I'm glad you're here.*

MATHISA: *I'm not sure I'm glad to be here. I've never been in this place before.*

DEAN: *It's scary to be in a place for the first time. We're always glad to welcome newcomers and visitors.(Smiles) What's going on ?*

MATHISA: *(Pauses) I'm here for a friend.*

DEAN: *Your friend is very lucky that you could come for him or her. How did you decide to come here?*

MATHISA: *Well, this is a place I pass every morning on my way to school. Sometimes I wonder what it's like here. And today I knew that I needed to come. Can I be sure that nobody will find out what I tell you?*

DEAN: *Yes, what you tell me stays between the two of us. Confidentiality is very important to you.*

MATHISA: *I have some information, and I don't want anyone else to know. I don't know what I can do.*

DEAN: *Umm . . . (Nods)*

MATHISA: *You need to know what before you can help, I guess.*

DEAN: *Could you describe the event that brought you here?*

MATHISA: *I'm just so scared and I don't know what to do.*

DEAN: *It's scary having information and not having any idea what to do with it. How do you think I can help you?*

MATHISA: *I don't know for sure. But I do know that you understand AIDS and you help people with AIDS. I only know what they taught us in school. (Mathisa is obviously in distress; she is almost in tears and is choosing her words carefully.)*

DEAN: *Your quiet voice and your tears let me know that the reason you came is very upsetting to you.*

MATHISA: *(Nods)*

DEAN: *(Silence)*

MATHISA: *My best friend just told me that she has AIDS. She got tested when she was on a trip a month ago. She went to a state that does not ask your real name. She just found out yesterday. She's really blown away by it. No one else knows—not even her parents.*

DEAN: *She told you and you don't know what to do.*

MATHISA: *I don't really know anything about it. I don't want her to die, and I don't want to die. Her boyfriend doesn't know, and I don't know what she'll tell her parents. She may even run away or kill herself, but what if the tests are wrong? And seeing the really sick people here makes me think that I don't want to live.*

DEAN: *Mathisa, I'm not sure what you said just then; you said that you didn't want your friend to die, and then you said that you didn't want to die.*

In this interview, Dean promoted good rapport with Mathisa by providing a good physical setting. It was simple, without distractions; they sat in close proximity, with no barriers between them, in comfortable chairs. Perhaps most important, it was an environment that was private. Dean introduced himself and assured her of confidentiality so that Mathisa felt comfortable beginning to talk.

Dean used a combination of open inquiries and responses. His first open inquiry was "Can you tell me why you're here?" This was designed to elicit a fact from Mathisa. She did not elaborate, but she did give enough information to continue the conversation. Dean also used "how" and "could" questions to encourage Mathisa to provide more information.

Dean's responses also included a paraphrase ("Confidentiality is very important to you") as well as reflection ("Your quiet voice and your tears let me know that the reason you came is very upsetting to you."). Both of these responses helped Mathisa understand that Dean was actively listening to her and had heard what she had said. He had also interpreted her nonverbal messages.

At the conclusion of this excerpt from the interview, Dean used clarification ("I'm not sure what you said just then . . .") to try to sort through the information that Mathisa has given. In the remainder of the interview, Dean will continue to find out more about the problem and its implications for Mathisa and her friend. When they finish talking, Dean will summarize what has transpired and perhaps suggest where the relationship can go at that point.

Interviewing Pitfalls

Clearly, interviewing requires a great deal of skill. An effective interviewer is one who listens attentively, questions carefully, and uses other helpful responses to elicit information and promote client understanding. However, caution is necessary. The desire to be helpful and the anxiety of conducting that first interview can lead to a number of pitfalls. Four of them will be discussed here.

Premature problem solving   This arises from a desire to be helpful to the applicant by removing the pain, the discomfort, or the problem itself as soon as possible. Unfortunately, if the interviewer suggests a change, strategy, or solution before the problem has been fully identified and explored, this may address a symptom of the presenting problem rather than the actual problem. Premature problem solving may cause the client to lose confidence in the case manager's knowledge and skills or to become impatient. Also, premature problem solving undermines the client's self-determination and can lead to false assumptions, misinterpretation of what the client says, and steering him or her in the wrong direction. In the case of mental illness, misdiagnosis can result.

Giving advice   In attempting to solve the problem or offer a solution, the case manager may mistakenly give advice. When given hurriedly and before the problem has been explored sufficiently, advice may be seen as indicating a lack of interest or thoroughness. The client may also feel misunderstood, or he or she may superficially agree, without intending to follow through. Advice giving also tends to diminish the client's level of responsibility, self-determination, and partnership in problem solving.

Overreliance on closed questions   The pitfall of overuse of closed questions has been discussed elsewhere in this chapter. Remember that closed inquiries are usually directive and focused on facts; they rarely provide the opportunity for exploration. A series of closed inquires may also make the client defensive. Once this feeling is established, it is difficult to overcome.

Rushing to fill silence   Because silence is often awkward in everyday social situations, beginning helpers as well as seasoned professionals are sometimes uncomfortable with pauses and rush to fill them, believing that silence indicates that nothing is happening. In fact, silence does have meaning. The client may be waiting for direction from the interviewer, thinking about what has transpired so far, or just experiencing an emotion. Constant dialogue can be a false signal that something is happening. Skillful case managers learn to listen to silence.

 # Chapter Summary

This chapter focuses on the interviewing process, especially the intake interview that initiates the helping process. Critical to this process are the helper's values and attitudes since these can convey to the client how the helper feels about him- or herself and how the helper feels about the client. A positive interaction is more likely to occur if the helper demonstrates warmth and caring. Attention to the physical space can also facilitate the intake interview process. For example, talking with the client in a private area can signal that confidentiality is important. Barriers that discourage the client can be a lack of sensitivity to racial, religious, cultural or gender issues.

Basic communication skills are important if the case manager is to establish a dialogue with the client. Effective communication includes demonstrating congruence between what is said verbally and nonverbally, engaging in active listening, and being sensitive to cultural differences. Listening is key to conducting an effective interview. This means imparting to the client that attention is being paid to what is said, which conveys respect and the desire to learn about a client. Listening also includes responsive listening and the active listening described earlier. Questioning is also a skill and an art. Used appropriately, questioning is an effective way of helping people talk about themselves without asking direct, closed questions.

Good interviewers not only develop communication skills, they also learn to avoid pitfalls, including premature problem solving and giving advice. Case managers are often tempted to identify the problem too quickly or to seek an immediate solution. Both of these responses may focus more on the case manager than the client. Another pitfall is an overreliance on closed questions, which discourage clients from talking. The final barrier is rushing to fill silence rather than giving clients time to assume responsibility for part of the dialogue. These barriers can be replaced with other communication skills that encourage client participation.

Chapter Review

◆ *Key Terms* ....

Sexism   *Questioning*
Racism   *Attending behavior*
Ethnocentrism   *Closed questions*
Ageism   *Open inquiries*
Effective communication skills   *Verbal following*
Active listening

◆ *Reviewing the Chapter* ....

1. What attitudes and characteristics facilitate the development of a helping relationship?
2. Write a dialogue representing the beginning of an intake interview to illustrate desirable attitudes and characteristics of the helper.
3. Describe an office setting that facilitates relationship building.

4. Discuss the problems that are created by stereotypes based on gender, race, and age.
5. Give general guidelines for essential communication skills in interviewing.
6. Discuss the importance of listening in the intake interview.
7. What is attending behavior (active listening)?
8. What are the five listening behaviors represented by the acronym S-O-L-E-R?
9. Why are listening and questioning both complex skills and arts?
10. State the advantages and disadvantages of questioning.
11. Describe the five situations in which questioning is appropriate.
12. Distinguish between closed and open inquiries.
13. What are the four commonly used methods of introducing an open inquiry?
14. Name four pitfalls of interviewing and tell how each may be avoided.

◆ *Questions for Discussion* ....

1. Do you think that you will be able to conduct a good interview? What skills will you need to strengthen your competence as an interviewer?
2. Speculate about how interviewing will change if computers are used in the process.
3. Discuss the kinds of activities that might help you practice your listening skills.
4. Do you believe that certain communication skills are essential to effective interviewing? If your answer is no, why not? If yes, what are they, and why are they important?

References

Baruth, L., & Manning, M. (1999). *Multicultural counseling and psychotherapy: A lifespan perspective*. Englewood Cliffs, NJ: Prentice Hall.

Bernal, G., Martinez, A. C., Santisteban, D., Bernal, M. E., & Olmedo, E. E. (1983). Hispanic mental health curriculum for psychology. In J. C. Chunn, P. J. Dunston, & R. Ross-Sheriff (Eds.), *Mental health and people of color* (pp. 65–96). Washington, DC: Howard University Press.

Brammer, L. M. (1993). *The helping relationship*. Englewood Cliffs, NJ: Prentice Hall.

Brown, D., & Srebalus, D. J. (1996). *Introduction to the counseling profession*. Boston: Allyn & Bacon.

Carkhuff, R. R. (1969). *Helping and human relations: A primer for lay and professional helpers. Vol 1: Selection and training*. New York: Holt, Rinehart & Winston.

Combs, A. W. (1969). *Florida studies in the helping professions*. Gainesville: University of Florida Press.

Egan, G. (1998). *The skilled helper: A systematic approach to effective helping*. Pacific Grove, CA: Brooks/Cole.

Eichinger, J., Rizzo, T. L., & Sirotnik, B. W. (1992). Attributes related to attitudes toward people with disabilities. *International Journal of Rehabilitation Research, 15*, 53–56.

Epstein, L. (1985). *Talking and listening: A guide to the helping interview*. St Louis: Times Mirror/Mosby.

Evans, D. R., Hearn, M. T., Uhlemann, M. R., & Ivey, A. E. (1998). *Essential interviewing: A programmed approach to effective communication.* Pacific Grove, CA: Brooks/Cole.

Gil, R. M., & Vasquez, C. I. (1996). *The Maria paradox: How Latinas can merge Old World traditions with New World self-esteem.* New York: Putnam.

Gilligan, C. (1982). *In a different voice.* Cambridge, MA: Harvard University Press.

Gordon, T. (1974). *Teacher effectiveness training.* New York: David McKay.

Ivey, A. E., & Ivey, M. B. (1999). *Intentional interviewing and counseling: Facilitating client development in a multicultural society.* Pacific Grove, CA: Brooks/Cole.

Kirst-Ashman, K. K. (1999). *Understanding generalist practice.* Belmont, CA: Wadsworth.

Kolanad, G. (1994). *Culture shock! India.* Singapore: Time Books International.

Long, L., Paradise, L. V., & Long, T. J. (1981). *Questioning: Skills for the helping process.* Pacific Grove, CA: Brooks/Cole.

Morales, A., & Sheafor, B. W. (2001). *Social work: A profession of many faces.* Boston: Allyn & Bacon.

Sielski, L. M. (1979). Understanding body language. *Personnel and Guidance, 57*(5), 238–242.

Smith, S. (1995a). Family theory and multicultural family studies. In B. B. Ingoldsby & S. Smith (Eds.), *Families in multicultural perspective* (pp. 5–35). New York: Guilford Press.

Smith, S. (1995b). Women and households in the Third World. In B. B. Ingoldsby & S. Smith (Eds.), *Families in multicultural perspective* (pp. 235–267). New York: Guilford Press.

Woodside, M., & McClam, T. (2002). *Introduction to human services.* Pacific Grove, CA: Brooks/Cole.

Yuker, H. E., (1994). Variables that influence attitudes toward people with disabilities: Conclusions from the data. *Journal of Social Behavior and Personality, 9*, 3–22.

Chapter Six

Service Delivery Planning

We call our clients "consumers" because they are here for our services. They are buying our services. There are other agencies doing exactly what we are doing, so they can certainly go to another agency that they find more amicable or friendlier. So they are definitely consumers.

     —Yolanda Vega, CSW, Director of Agency Services, Casita Maria Settlement House, Bronx, NY, personal communication, May 4, 1994

The treatment plan is developed by administering the Millon Adolescent Clinical Inventory to each new resident. It is a 160-item questionnaire that covers a broad range of issues under three headings that include personality patterns, expressed concerns, and clinical syndromes. Ones the questionnaire is scored and the issues of concern are established, the resident and case manager develop a treatment plan around those issues.

     —Stacie Newberry, Marion Hall Emergency Shelter, St. Louis, MO, personal communication, December 13, 1999

In the interview, I tell clients what we expect in our program. For example, they have to have a 9.5 in math and 10.0 in reading on their TABE test. They have to be able to type at least 30 words a minute. They know their expectations at the very beginning. And they are only allowed to miss up to 8 days in this training program. . . .

     —Cathy Lowe, Private Industry Council, Knoxville, TN, personal communication, August 30, 1993

At this point in the process, the agency has determined that the applicant meets the eligibility criteria, the services are appropriate, and the person can now receive services. At Casita Maria Settlement House, the applicant becomes a "consumer," whereas the Private Industry Council uses the term *client*. An agency in South Dakota that serves adults with developmental disabilities calls the service recipients *individuals*, explaining that "they are not clients anymore or consumers. They are just people." Other agencies or organizations use the term *customer*. The change in status from applicant to recipient of services marks the move into the second phase of case management: planning service delivery.

    The quotes that introduce this chapter identify some of the activities that occur during this phase. At Marion Hall, client participation is important in planning. In fact, clients determine the goals. Client interests and expectations and use of test data are shared at the Private Industry Council. Margaret Mikol at SKIP in New York City summarizes this phase of case management: "You have to sort through the information and say, 'Now, what are the real issues here?' Then finally you have to be able to say, 'Okay, if this is the real problem, what do I have to do to solve it?'" (personal communication, May 4, 1994).

Linda Smith, a caseworker at the School for the Deaf, also notes the importance of gathering information in order to see the big picture: "Probably one of the most important things in my job is being able to communicate with all the people who are involved with the student because . . . so many people have different pieces of information, and if we can pull all the information together, we can get a better picture of the way to work with our kids or their families. I think being able to see the whole in addition to the individual little narrow parts is a very important characteristic for somebody to be able to see that big picture" (personal communication, October 27, 1993).

This chapter explores the planning phase of case management, wherein the helper and the client together determine the steps necessary to reach the desired goal. The activities involved in this phase include the review and continuing assessment of the problem, the development of a plan, the use of an information system, and the gathering of additional information. Running through our discussion in this chapter are two critical components of the case management process—client participation and documentation.

For each section of the chapter, you should be able to accomplish the objectives listed here.

REVISITING THE ASSESSMENT PHASE
• List the two areas of concern that are addressed when reviewing the problem.

DEVELOPING A PLAN FOR SERVICES
• Identify the parts of a plan.
• Write a plan.

IDENTIFYING SERVICES
• Locate available services.
• Create an information and referral system.

GATHERING ADDITIONAL INFORMATION
• Compare interviewing and testing as data collection methods.
• Identify the types of interviews.
• Show how sources of error can influence an interview.
• Illustrate the role of testing in case management.
• Define *test*.
• Categorize a test.
• Identify sources of information about tests and the information that each provides.
• Analyze the factors to be considered when selecting, administering, and interpreting a test.

Revisiting the Assessment Phase

The next phase of case management begins with a review of the problems identified during the assessment phase. Before moving ahead with the process, the case manager will need to know if the problem has changed, if the same client resources are available, and if any shift in agency priorities has occurred. According to Janelle Stueck at the Private Industry Council, "one of the most difficult pieces of case management is doing that continual assessment and not resting on what you think things were two weeks ago. They may be different" (personal communication, August 30, 1993). In order to complete the review quickly before moving into a planning mode, the case manager and the client examine two aspects of a case.

The first area of concern involves a review of the relevant facts regarding the problem. At this point, the case manager and the client revisit the identification of the problem. The initial question the helper asks can help determine whether the problem still exists. Working with people requires an element of flexibility; clients' lives change, just as ours do. Thus, the problem may have changed in some way, the client may have a different perspective on it, the participants may be different, or assistance may no longer be needed, or appropriate, or wanted. Once the case manager has confirmed that the problem still exists and has documented any changes that have occurred, the problem itself is revisited. Is the problem an unmet need, such as housing or financial assistance, or is it stress that limits the client's coping abilities or causes interpersonal difficulties? Is the problem a combination of several factors? This activity is best accomplished by talking with the client and reviewing his or her file. The client is still considered the primary source of information and a partner in the case management process.

A second area of concern in the review of the problem requires an examination of available information to answer the following five questions.

- What do I know about the source of the problem?
- What attempts have been made previously (before agency contact) to resolve the problem?
- What are the motivations for the client to solve the problem?
- What are the interests and strengths of the client that will support the helping process?
- What barriers may affect the client's attempts to resolve the problem?

An important source of information is the client. Talking with the client can reveal what he or she has thought about doing, what has been tried, and some possible solutions. Exploring with the client his or her motivations, strengths, and interests indicates that the process of case management continues to be a partnership between the client and the case manager.

Other techniques that are helpful in reviewing the problem are observations and documentation. In the course of receiving the application, conducting the intake interview, making a home visit, or all three, the case manager has had opportunities to observe the client. These observations may be richer if they occur in the home or in the office or if the client is accompanied by family members or a significant other. Information available from such observations includes the client's thoughts, feelings, behaviors, and relationships.

Documentation in the case file also provides facts and insights about the client. Case notes, reports from other professionals, and intake forms help the case manager pin down past occurrences and pertinent facts about the present situation. Case managers who have a long history in service delivery may call on knowledge and experience from the past to understand a current case. Sometimes, knowledge comes from a case manager's "own perception, instinct, kind of experience, street know-how" (McClam & Woodside, 1994, p. 41). Many case managers mention rapid insight they sometimes have about a client, the client's environment, possible difficulties, and creative approaches to the case management process. This insight is treated as just one piece of information and must undergo the same scrutiny as the other information collected.

Once the case manager has revisited the problem, confirmed its existence, documented any changes, and reaffirmed the client's desire for assistance, the two of them move to the next step of the planning phase, which addresses the need to determine the steps necessary to reach the identified goal or goals. This is the plan that will guide service provision.

Developing a Plan for Services

The **plan** is a document, written in advance of service delivery, that sets forth the goals and objectives of service delivery and directs the activities necessary to reach them. The plan also serves as a justification for services by showing that they meet the identified needs and will lead to desired outcomes. More specifically, a plan describes the service to be provided, who will be responsible for its provision, and when service delivery will occur. If there are financial considerations, the plan may also identify who will be responsible for payment. Sometimes financial support is available from outside sources, including the client and the family. Usually, the completed plan is signed by the client and the case manager as the representative of the agency. It may then be approved by someone else in the agency before the authorization to provide services is granted.

Clearly, the plan is a critical document, since it identifies needed services and guides their provision. How is it developed? What is included? What are goals and objectives? These questions are answered as you read this section.

Plan development is a process that includes setting goals, deciding on objectives, and determining specific interventions. The process begins with the synthesis of all the available data. This information is scrutinized carefully for as complete a picture of the case as possible. It is analyzed so as to identify desirable outcomes.

For the beginning case manager, the following method uses a step-by-step approach to synthesize data and integrate the information into a workable plan. Using the worksheet displayed in Table 6.1, the case manager can record his or her analysis.

TABLE 6.1 INTEGRATING CLIENT INFORMATION

CLIENT WORKSHEET

Client Name: _____

Date: _____

Source of Information	Relevant Facts	Conclusions	Contradictions	Missing Information	Motivations of Client	Strengths of Client	Interests of Client

- Reread the client file and fill in the following categories on the worksheet: sources of information, relevant data.
- With this snapshot of the contents of the client's file, assess and record conclusions, contradictions, and missing information.
- Review this assessment with the client and make revisions according to his or her input and other new data gathered; fill in client motivations, interests, and strengths with client input.
- Discuss with the client desirable outcomes.

In Roy Roger Johnson's case in Chapter 1, the information available at the time of plan development was derived from Roy's application for services, the intake interview, reports from his orthopedic surgeon, case documentation, a general medical examination report, a psychological evaluation, and a vocational evaluation report. When Roy and his counselor developed the plan of services, they reviewed and considered all this information using the steps listed in Table 6.1.

Roy had a back injury and needed assistance finding a job; he also met economic eligibility criteria. His service plan, reproduced in Figure 6.1, included a program objective and intermediate objectives. For each objective, a service was identified, as well as a method of checking progress toward the achievement of the objective. The form also provided space to describe any other client, family, or agency responsibilities or conditions. Because this agency values client participation, Roy's view of the program was also noted. Then both Roy and the counselor signed the plan.

Exactly what a plan looks like varies from agency to agency. However, if you are employed by an agency that provides case management or client services, you can be sure that a plan will guide your work. Let's examine the components of a plan of services.

Service plans are goal directed and time limited, so they should include both long-term and short-term goals. Long-term goals state the situation's ultimately desired state. Short-term goals aim to help the client through a crisis or some other present need. Whatever the time constraints, goals establish the direction for the plan and provide structure for evaluating it.

Goals are statements that describe a state or condition or an intent. For clients, a goal is a brief statement of intent concerning where they want to be at the end of the process; for example, "Learn daily living skills in order to live independently," "Acquire knowledge and skills for a career in business communications," or "Develop a support network for help coping with phobias."

Having written goals helps us focus on what we are trying to accomplish before we take action or provide any services. Action is often easy, but sometimes relating actions to outcomes is not. For accountability reasons, service provision is tied to outcomes. This makes writing goals a critical step in plan development. Remember that these broad statements of intent can be achieved only to the degree that their meaning is understood, so well-stated, reasonable goals are essential to problem resolution.

At Intensive Case Management in Los Angeles, clients decide on their goals.

SERVICE PLAN

1. NAME __JOHNSON, ROY__ — PROGRAM TYPE ☒ INITIAL ☐ AMENDMENT

2. YOU ARE ELIGIBLE FOR: ☒ VOCATIONAL REHABILITATION SERVICES ☐ EXTENDED EVALUATION SERVICES ☐ PAST EMPLOYMENT SERVICES

   BECAUSE: ☒ A. YOU HAVE A PHYSICAL OR MENTAL DISABILITY WHICH CONSTITUTES A SUBSTANTIAL HANDICAP TO EMPLOYMENT AND:

   ☒ B. YOU CAN REASONABLY BE EXPECTED TO BENEFIT IN TERMS OF EMPLOYABILITY FROM SERVICES.

   ☐ C. IT CANNOT BE DETERMINED WHETHER OR NOT YOU CAN BENEFIT IN TERMS OF EMPLOYABILITY FROM REHABILITATION SERVICES.

   ☐ D. POST EMPLOYMENT SERVICES ARE NEEDED FOR YOU TO MAINTAIN EMPLOYMENT.

3. PROGRAM OBJECTIVE: __Business Communications__ ANTICIPATED DATE OF ACHIEVEMENT: MONTH _1_ YEAR _XX_

4. INTERMEDIATE OBJECTIVE, SERVICES METHODS OF CHECKING PROGRESS.

   ESTIMATED DATES TO REACH OBJECTIVE & RECEIVE SERVICES

	RESPONSIBILITY	FROM	TO
OBJECTIVE To correct physical impairment so that client might			
SERVICES reach vocational objective	Client	1/XX	1/XX
Possible office visit with the doctor			

METHOD OF CHECKING PROGRESS __Medical information__

	RESPONSIBILITY	FROM	TO
OBJECTIVE To provide background information and educational skills			
SERVICES so that client might reach vocational objective	VR	1/XX	1/XX
A. Tuition/UT Knoxville		1/XX	1/XX
B. Miscellaneous Educational Expenditures		1/XX	1/XX

METHOD OF CHECKING PROGRESS __R-11, Grade Reports__

	RESPONSIBILITY	FROM	TO
OBJECTIVE To follow client's progress and develop plan amendment if needed so that client might			
SERVICES reach objective	Client	1/XX	1/XX
A. Possible RP-B		1/XX	1/XX
B. Client/Counselor Contacts		1/XX	1/XX

METHOD OF CHECKING PROGRESS __R-11__

5. CLIENT OR FAMILY AND AGENCY RESPONSIBILITIES AND CONDITIONS: I. Client is responsible to maintain contact with counselor twice each semester by mail, phone or in person. II. Client is responsible to furnish VR Counselor with a copy of grades at the end of each term. III. Client is responsible to maintain an average load of classes and average grades throughout his program. IV. Client is responsible to file for any similar benefits which might help him pay for his program. V. Client is responsible to furnish VR counselor with a resume and a list of potential employers to interview with during the first part of his senior year. VI. Client is responsible to notify counselor of any significant change of address, health, phones number of financial status.

6. CLIENT'S VIEW OF PROGRAM The client and I have discussed the services necessary to help him reach his vocational objective and we are in mutual agreement with his plan.

I HAVE PARTICIPATED IN THE DEVELOPMENT OF THIS PROGRAM AND I UNDERSTAND IT.
I UNDERSTAND AND ACCEPT THE STATEMENT OF UNDERSTANDING WHICH HAS BEEN EXPLAINED TO ME.

Roy Johnson — Client's Signature   5-6-XX Date   *Susan Fields* Supervisor Signature   5/6/XX Date

Figure 6.1   Roy's service plan

"We put it in their words. Then we identify with the client some of the barriers to reaching the goal. What is going to keep you from doing this? Then we talk about what we are going to do. What is a little goal that we can accomplish in the next six months? Sometimes they have a hard time dealing with little goals. So we write it down and list some of the steps. We also decide what the roles of family, friends, staff, or others will be" (Jan Cabrera, personal communication, March 24, 1998).

How does one write goals that are well stated and reasonable? Three criteria help us achieve this. First, the goal should be expressed in language that is clear and concise; second, the goal statement should be unambiguous; and third, the goal must be realistic and achievable. These criteria are illustrated in the following goals, which were established for a 74-year-old woman who will attend the Daily Living Program at the Oakes Senior Citizens Center.

Draft 1 is a *goal statement* for Ms. Merriweather; Draft 2 improves the statement by making it more clear and concise.

> *Draft 1:* Ms. Merriweather will participate often in many of the Oakes programs that relate to sports, games, music, communication, exploring other cultures, and other educational programs as they are developed by the creative staff in the activities area.
> *Draft 2:* Ms. Merriweather will increase her social opportunities by participating in center activities.

A description of the *plan* is presented in Draft 1, below. In Draft 2, it is restated less ambiguously by defining who will help with medications and what the help entails.

> *Draft 1:* They will work with Ms. Merriweather and her numerous family members to help with medications.
> *Draft 2:* Nursing staff will develop a plan to administer Ms. Merriweather's medication.

The goal in Draft 1, below, establish general physical goals for Ms. Merriweather. Draft 2 restates these goals in realistic and achievable terms.

> *Draft 1:* Ms. Merriweather will increase her range of motion, physical strength, and stamina.
> *Draft 2:* Ms. Merriweather will participate four times a day in an exercise program that includes walking, weightlifting, and stretching.

Thus, goals are an important part of the service plan. They increase the chance of solving the problem by providing direction and focusing attention on well-expressed, reasonable statements. Since formulating goals also requires collaboration between the client and the case manager, writing them also highlights the shared responsibility for the case. Once a broad statement of intent has been agreed on, it is time to identify the activities that will lead to the desired outcomes. This process continues as a cooperative effort between the client and the case manager. Activities are identified as objectives.

An **objective** is an intended result of service provision rather than the service itself. It tells us about the nuts and bolts of the plan—what the person will be able to do, under what conditions the action will occur, and the criteria for acceptable performance—so that we can know whether the objective has been accomplished. Objectives are useful for several reasons. First, they tell us where we are going. Second, they give the client guidance in organizing his or her efforts by stating the intervention or action steps. Third, they state the criteria for acceptable performance or outcome measures, thereby making evaluation possible. Objectives are all-important for the case manager since they provide the standards by which progress is monitored. As progress is made, the case manager adjusts the plan as needed.

Writing clearly defined objectives benefits the client, the case manager, and the agency. Boserup and Gouge (1977, p. 111) provide the following guidelines for writing and evaluating service objectives:

1. The statement of objective should begin with the word *to* followed by an action verb. The achievement of an objective must come as a result of action of some sort. Therefore, the commitment to action is basic to the formulation of an objective.

2. The objective should specify a single key result to be accomplished. For an objective to be effectively measured, there must be a clear picture of when it has or has not been achieved.

3. The objective should specify a target date for its accomplishment. It is fairly obvious that to be measurable, an objective must include a specific completion date, either stated or implied. If the objective is of a continuing nature, the target date could be assumed to be the end of the eligibility period. A situation of this nature may occur when services are being provided to a client whose prospects for improvement seem very slim.

4. An objective should specify the *what* and *when*; it should avoid venturing into *why* and *how*. Once again, an objective is a statement of results to be achieved. The "why bridge" should have been crossed before the actual writing of the objective has started. The means of achieving an objective should not be included in the objective statement.

5. Objectives should be realistic and attainable but still represent a significant challenge. Since an objective can and should serve as a strong motivational tool for the individual worker and client, it must be one that is within reach. This simply means that resources must be available to achieve the objective.

6. Objectives should be recorded in writing. Each of us, whether consciously or unconsciously, has a convenient memory: We tend to remember the things that turn out the way we want them to and either forget or modify those things that are less than we wish. If objectives were not put in writing, it would be relatively easy to look on accomplishments as if they were in fact planned objectives. On the other side of the coin, one of the sharpest areas of conflict among case manager, client, and supervisor is illustrated by such phrases as "I thought you were working on something else!" or "That's not what we agreed to do" or "You didn't tell me that's what you expected." Having objectives in writing will not eliminate all these problems, but it will provide something more tangible for comparison. Furthermore, written objectives serve as a constant reminder and an

effective tracking device by which the case manager, the client, and the supervisor can measure progress.

7. A statement of objective must be consistent with the available or anticipated resources.

8. Ideally, an objective should avoid or minimize dual accountability for achievement when joint effort is required.

9. Objectives must be consistent with basic agency policies and practices.

10. The client must willingly agree to the objectives without undue pressure or coercion.

11. The setting of an objective must be communicated not only in writing but also in face-to-face discussions with the client and the resource persons or agencies contributing to its attainment.

The following case example illustrates the development of goals and objectives (including intervention and outcome measures) with a client who is elderly and needs assistance.

Mary Sue Davis is an 86-year-old white married female. Her husband has recently been placed in a nursing home facility so that he can receive full-time care. Mrs. Davis has a severe heart condition and has been ordered by her physician to rest every 2 hours and not to travel by herself because of dizzy spells. The nursing home is now receiving her husband's Social Security income. Mrs. Davis lives in a two-bedroom apartment. They have one son who lives an hour away and also has a heart condition. Mrs. Davis is requesting assistance with transportation in order to visit her husband on a more regular basis.

An interview with Mrs. Davis at her apartment revealed that her income consists solely of her Social Security checks. She does have Medicare to help with the costs of treatment for her heart condition. She currently uses public transportation (bus) to travel where she needs to go. During the interview, the service coordinator identified additional problems: the affordability of her current apartment, the availability of affordable housing, the need for an escort for travel, and possible grief issues regarding her husband's condition and placement in a nursing home.

Mrs. Davis agrees that she cannot afford her apartment and needs to seek more affordable housing. She is willing to apply for Community Action Committee (CAC) transportation that will pick her up at her door. She is very realistic regarding her husband's condition. Although she wishes he could come home, she has accepted that he will most likely remain at the nursing home. She realizes that she has to take care of her own health, but at the same time she has to get things done, and there is not always somebody around to help.

COUNCIL ON AGING
CLIENT PLAN

CLIENT Mary Sue Davis

DATE 11/6/XX

GOAL 1 Locate affordable housing

    Objective 1: To contact the city housing authority this week for application for rent-controlled apt.—service coordinator.

    Objective 2: To review list of apts., decide which ones to see, and select one (1 month)—Mrs. Davis.

    Objective 3: To request volunteer assistance to escort Mrs. Davis on apartment visits and to help with the move (1 week)— service coordinator.

GOAL 2 Provide transportation

    Objective 1: To complete application for K-Trans lift and CAC vans (this week)—Mrs. Davis.

    Objective 2: To determine eligibility for medical escort service (2 weeks)—service coordinator.

    Objective 3:

(continues)

Figure 6.2 Client plan for Mrs. Davis

The service coordinator identified two main goals for Mrs. Davis: to find affordable housing and to secure transportation that is appropriate. These are set forth in the Client Plan (Figure 6.2).

The first objective toward the housing goal was to complete an application for a rent-controlled apartment with the city housing authority. Due to long waiting

GOAL 3 _____

    Objective 1: _____
              _____
              _____

    Objective 2: _____
              _____
              _____

    Objective 3: _____
              _____
              _____

GOAL 4 _____

    Objective 1: _____
              _____
              _____

    Objective 2: _____
              _____
              _____

    Objective 3: _____
              _____
              _____

Service Coordinator _____   Client _____
                SIGNATURE   SIGNATURE

Figure 6.2   (*Continued*)

lists, this needed to be done within the week. The next step was to determine where she preferred to live (probably close to the nursing home). After the application was completed, the service coordinator arranged for a volunteer to take Mrs. Davis to look at several apartments and to meet with apartment managers to find out about waiting lists (Mrs. Davis couldn't afford to wait for long). The service coordinator found a volunteer to help with this. Once Mrs. Davis decided on an apartment, other volunteers assisted with the move. Her son could afford to rent a moving truck and to drive the truck, although he couldn't lift or carry due to medical problems. The time allotted for these objectives was workable, and the objectives were met within a month.

The objectives for the goal of transportation were to apply for the K-Trans lift along with CAC vans. Obtaining an assessment from the Office on Aging was also an objective; that agency provided escorted transportation for medical appointments and necessary errands for people over 60. This service would be available until Mrs. Davis was accepted by another agency that provides transportation.

In this case, the plan identified services and then guided the delivery of those services. The goals and objectives in the plan were developed using the guidelines suggested previously. Note that each objective clearly defined the intervention or action steps, stated who would provide the service, and stated a time frame for service delivery. The outcome measures were clear and the plan was implemented successfully.

 # Identifying Services

Once the plan is complete and has been agreed on by the client and the case manager, it is time to begin thinking about the delivery of services. A well-developed plan provides information about what the service is, who will provide it, what the time frame is, and who has overall responsibility for service delivery. It is the case manager's responsibility to implement the plan. What are these responsibilities? How does one begin implementation? These questions are explored next.

Identifying services has been compared to the brokering role. In both situations, the case manager is involved in the legwork and planning that is necessary for implementation. As a broker, the case manager helps clients access existing services and helps other service providers relate better to clients. This linking of clients and services also occurs as the case manager arranges for service delivery. The steps are similar.

Information and Referral Systems

One of the most helpful tools for a case manager is knowledge of the human service delivery system in the community. Who do you know? What services are available? How does one access the services? Is there a waiting list? One of the challenges facing new case managers is to establish an **information and referral system**. For case managers with experience, the challenge consists of continually developing and updating their systems. Knowing what an information and referral system is, how to set one up, and how to use it are valuable skills in case management.

Human service employers believe that the people their agencies will serve in the future will be multiproblem clients, such as people with dual diagnoses, diverse problems, and problems of long standing (McClam, 1992). The needs of clients such as these rarely match the services available from a single agency. In these cases, the case manager finds it invaluable to have information about other available services. Many helping professionals have personal service directories to supplement existing community or agency directories.

There are three components to information and referral. One component is the **social service directory**, which usually lists the kinds of problems handled and the services delivered by other agencies. In some communities, these are published by a social service agency, by a funding source such as the United Way, or (as a community service) by a business or organization. Sometimes these directories are available on the World Wide Web. Another component is the **feedback log**. Such logs provide feedback to the agencies that deliver services to help ensure quality information and referral services. Some agencies accomplish this through referral forms that record referrals, give information on the services needed, and provide the referral agency with information on the services received. If the client takes the form to the agency providing services, it may also serve to remind the client of the appointment. A third component of information and referral systems is staff training. In these sessions, the helper may be introduced to the services of the employing agency as well as those of other agencies. Other information and referral data that are shared during staff training may include reviewing and updating referral procedures, announcing new services or ones that no longer exist, and discussing the effectiveness and efficiency of service delivery.

Social service directories may have two indices: one that is an alphabetical listing of agencies and one that is a categorical listing of services. Each entry in the directory lists the agency's name, address, phone number, and services. Also listed may be fees, hours of service, eligibility criteria, and sources of agency support. Here is an example of an entry.

RUNAWAY SHELTER
2535 Magnolia Avenue
Bluff City, NJ

Purpose and services: The Runaway Shelter and Homeless Youth Shelter provides shelter, counseling, and casework services to youth (ages 13–18). This is a short-term (14 days) service.

Eligibility: No eligibility requirements

Fees: None

Hours open: 24 hours

Area served: Region

Sources of support: United Way, Department of Health and Human Services

Existing directories are helpful to the case manager, but sometimes establishing one's own system is useful for filling in the gaps in published directories or for recording detailed information that may be of special interest to the individual helper.

Setting Up a System

The first step in establishing one's own information and referral system is to identify all agencies and available services. This includes listing agencies previously contacted, checking the Yellow Pages of the telephone book, browsing the World Wide Web, and talking with other professionals. Each agency and service becomes part of a card file or a computer file that is easy to update. The file can also be expanded by talking with clients (particularly those who have been in the human service system for some time), meeting other professionals at meetings and workshops, and attending community meetings.

Whether on cards or in the computer, such a file is easy to use when identifying the client problem and matching it with a service. However, since a client rarely has only one problem, using the file may not be so simple. First, the client and the case manager prioritize the problems. Once this has been done, the case manager identifies which problems the agency will address and which ones need referral. These additional services can be found by checking the file. If there is more than one resource to serve the client's particular need, the case manager works to identify the agency that can meet the client's needs in a manner responsive to the client's values and concerns.

Deborah Caudill is an 18-year-old client who needs long-term counseling to work on the anger she feels toward her father for deserting the family when she was 11. Lou Levine, her case manager, knows that the counseling Deborah needs is beyond the scope of the services provided by the agency where she works. Two other agencies in their community offer long-term counseling for adolescents. Since Deborah and Lou agree that counseling would be beneficial, they discuss these two agencies. Deborah has questions about their locations, who provides the counseling, whether it is group or individual, and how much it will cost. Lou consults her file for the answers to these questions and provides Deborah with the information, and then they discuss the pros and cons of each option. The case manager's file indicates that one center provides counseling services and is well known for its work with adolescents. In addition, the latest entry in the file indicates that Jane Barkley, a previous client, had a positive experience there.

Establishing and using an information and referral system requires certain skills of the case manager. Being able to identify the client's problem, the community resources available to solve it, and the viable alternatives are all critical to the success of the system. Choosing a resource or a service requires the client's participation. The client may actually have the final say in the selection of the agency or service; the more accurate and complete the information about the agency, the better the decision will be. Finally, good research skills are helpful because the case manager continually works to locate potential community resource alternatives and to update data on existing agencies and services.

Part of the development of a plan is identifying services to meet the client's needs. The development of an information and referral system is useful here. Throughout plan development, data gathering continues to take place.

Gathering Additional Information

Gathering additional information may be part of the planning process or part of the plan itself. To decide whether additional information is necessary, there must be a review of available information from other agencies, the referral source, employers, and others. The key to determining what is needed is relevance. Is the needed information relevant to the client and to service provision? Will it contribute to a complete array of social, medical, psychological, vocational, and educational information about the client? Once it is determined that additional information is necessary, the case manager decides how the information will be obtained. In some cases, the case manager can personally acquire the information, but it may also be necessary to consult family members, a significant other, or professionals such as psychologists, physicians, and social workers. The client also continues to be a primary source of information and is part of the decision-making process regarding the additional information needed and who can provide it. Next we introduce two data collection methods that case managers use; Chapter 7 explores what data are available from other professionals.

Data Collection Methods for the Case Manager

Two primary tools are available to the case manager for data collection: interviewing and testing. They are similar in several ways. The information is used to describe, to make predictions, or both. Each may occur in an individual or group situation in which some type of interaction occurs. The group situation may be an interview with a family or a test administered to more than one examinee. Both interviews and testing have a definite purpose, and the case manager assumes responsibility for conducting the interview or administering the test.

Interviewing

There are different types of interviews (Kaplan & Saccuzzo, 1997). The **assessment interview** is an interaction that provides information for the evaluation of an individual. The interview may be structured or unstructured; it uses both open-ended and closed questions. The intake interview is an example of an assessment interview in which the applicant provides information that helps in evaluating him or her and the problem in relation to the mission, resources, and eligibility criteria of the agency.

A **structured clinical interview** consists of specific questions, asked in a designated order. This type of interview is structured by guidelines to ensure that all clients are handled in the same way. The structure also makes it possible to score the responses. One advantage of this type of interview is its reliability or consistency. Flexibility is limited. Although it is a valuable source of information, the interview results should be interpreted with caution. The major limitation is its reliance on the respondent as an honest and capable interviewee who has skills for self-observation and insight.

A more comprehensive interview is the **case history interview**. This interaction includes both open-ended questions and specific questions. Topics may include a chronology of major events, the family history, work history, and medical history. Usually an interview of this type begins with an open-ended question or statement: *"What was school like for you?" "Tell me about your work history." "What do you remember as the happiest times when you were growing up?" "Describe your relationship with your parents."* These probes may be followed by specific questions, which may or may not be dictated by agency forms or guidelines. *"When did you quit your last job?" "What grade did you complete in school?" "Are you the oldest child?"* These questions help the case manager understand the client's background and uncover any pertinent information.

Technology is also an influence on interviewing. A computerized interview takes place via computer rather than face to face. Questions are presented and followed by a choice of responses.

Are you married?   Yes   No

If the answer is yes, then another question related to marriage may follow.

Is this your first marriage?   Yes   No

If the answer to the first question is no, then another question appears.

Did you complete high school?   Yes   No

The computerized interview is a good way to collect facts about a person. The limitations are that there is no nonverbal communication, and the feelings of the client are not shared. Important information may be lost as a result of these limitations.

The *mental status examination* (see Chapter 4) is a special type of interview used to diagnose psychosis, brain damage, and other major mental health problems. The purpose of a mental status examination is to evaluate a person thought to have problems in terms of factors related to these problems. The interview focuses on appearance, attitudes, behavior, emotions, intelligence, attention, and sensory factors. This type of interview requires the case manager to have some expertise on major mental disorders and the various forms of brain damage.

The skillful interviewer also needs to know about **sources of error** in the interview. Awareness of sources of potential bias in the instrument itself or in the interviewer enables the case manager to compensate for any resulting distortions. A look at interview validity and reliability will help us identify potential sources of error.

For a number of reasons it is often difficult to make accurate, logical observations and judgments. One is the **halo effect** (Thorndike, 1920), which occurs in an interview situation when the interviewer forms a favorable or unfavorable early impression of the other person, which then biases the remainder of the judgment process. For example, an unfavorable initial impression can make it difficult to see positive aspects of a client or a case. If a home visit to an apartment in a housing project reveals an unkempt, dirty, and very sparsely furnished living area, the case manager may find the visit unpleasant. The resulting interview with the single-parent resident is likely to be rushed and cursory, with little chance of gaining insight into any problems. The case manager may also find it difficult to maintain eye contact with the parent, thereby missing important nonverbal cues. Other contacts with this parent may be influenced by the memory of the physical setting.

A second cause of invalidity in an interview is "general standoutishness" (Hollingsworth, 1922). This is the tendency to judge on the basis of one outstanding characteristic, such as personal appearance. A more attractive, well-groomed individual might be rated more intelligent than a less attractive, unkempt individual. Consider a case manager who makes a home visit to investigate a child abuse report. The address is in an affluent suburb and the house is a stately two-story brick house with elaborate landscaping. The initial impression of neatness, money, and social standing may influence the investigator's interaction with the parents and the subsequent course of the investigation.

Cultural differences can also contribute to error. To take an extreme example, a case manager has been asked to visit a family that recently immigrated from India and has just moved into a rent-controlled apartment in the city. It is her last stop of the day, and she finds that she has interrupted a ceremony of *puja* (prayers of thanks for their new home). She finds family members seated on the floor around a fire. Appalled that they have started a fire on the floor, she stamps it out and begins lecturing the family on fire safety. When she finally begins to talk about the services that are available, the family does not respond.

As you can see, sources of error can prejudice interview validity. Error reduces the objectivity of the interviewer, often leading to inaccurate judgments. The more structured the interview is (see Chapter 4), the less error there will be. Because an interview does provide important information, the case manager can consider the information tentative and seek confirmation from other sources, such as more standardized procedures. Similarly, test results are more meaningful if placed in the context of a case or social history or other interview data. The two can complement each other.

The reliability of an interview is its consistency of results. In interviewing, this means that there is agreement between two or more interviewers in their conduct of the interview, the questions they ask, and the responses they make. As you might imagine, reliability varies widely. The reliability of structured interviews is higher because they have more stringent guidelines concerning the questions and even the order of the questions. (The downside is that this structure limits what is obtained.) In general, interview data have limited reliability, because interviewers look for different things, have various interviewing styles, and ask different questions. It is important for the case manager to verify information with other sources over time.

Testing

In the previous section, testing was recommended as one way to verify the information gathered in an interview. Most people encounter tests shortly after beginning school. How we perform on tests affects our lives, and test scores have become key factors in many decisions. They influence placement in special academic classes; the assignment of labels such as *high achiever*, *mentally challenged*, and *average*; admission to schools and colleges; and job selection. In fact, test scores are more important today than ever before.

Case managers encounter tests in various contexts; for example, test reports from other professionals. In some cases, the information consists of test scores and nothing more. Figure 6.3 shows one example of how test results may be communicated.

In other cases, test scores are part of a written report that also gives some explanation of the scores. Box 6.1 is an excerpt from a report on a 37-year-old white male who was hospitalized for depression. He has completed two years of college and has been a personnel interviewer for ten years. To use this information, the reader of the report must have knowledge of tests and an understanding of test data.

A case manager may also encounter testing as a service offered by an agency. For example, a statewide evaluation facility located on the campus of a school offers services that include achievement testing for placement at the school and vocational testing for career development. Workers at the evaluation facility administer these tests to each client who is referred to the facility. Scores are interpreted and included in their evaluation reports, which are sent to the referring counselor. There may be other situations in which a knowledge of testing is important. For example, a case manager may be asked to select tests to be administered as part of the services required in a plan. This task requires knowing the sources of information about tests, the criteria for selecting a test, and eligibility for purchase and use. Such knowledge is also important when the case manager encounters a situation like the following.

A family on my caseload had trouble understanding the results of a recent assessment test that was administered at their son's elementary school. The school counselor who originally explained the results of the test used terms unfamiliar to the parents and did not answer the questions they asked. The parents feel that if they understood the results of the test, they could help their son in the areas where he was weakest. The parents have asked me to look at the test results and explain them again.

The case manager needs an appropriate level of testing knowledge in order to use tests as a resource. Because tests have assumed such importance today,

---

TEST DATA

Name _Joe Brown_   Date _X/X/XX_

Date of Birth _X/X/XX_   Age _26_   Sex: M F   Race: W B   Other _____

Marital Status: M S D W   Education _11th grade_

    Wide Range Achievement Test (WRAT)

Reading grade	_13.2_	SS	_116_	%ile	_82_
Spelling grade	_11.2_	SS	_106_	%ile	_66_
Arithmetic grade	_8.5_	SS	_93_	%ile	_32_

    Wechsler Adult Intelligence Scale

        Verbal IQ _119_ Performance IQ _106_ Full Scale IQ _114_

    General Clerical Test

Clerical subscore	_26_	%ile	_10_
Numerical subscore	_22_	%ile	_35_
Verbal subscore	_56_	%ile	_40_
Total score	_104_	%ile	_25_

    Other tests:

Figure 6.3   Test data

particularly in decision making, case managers must think carefully about the role of testing in their work with clients. To make proper use of test results, one must understand the test being used; the purpose of the test, its development, its reliability and validity, administration and scoring procedures, the characteristics of the norm groups, and its limitations and strengths. This section presents an overview of these areas.

BOX 6.1   Test Administered and Results
● ● ● ● ● ● ● ● ● ● ● ●

Wide Range Achievement Test: A measuring device to estimate grade levels in three academic areas. Results from administering the WRAT show that the client is functioning at the 17.4 grade level in reading, 15.0 in spelling, and 7.1 in arithmetic. These scores are within the very superior, superior, and average classifications, respectively, when compared with the appropriate normative age group.

What is a test?   A **test** is a measurement device. A **psychological test** is a device for measuring characteristics that pertain to behavior. It is a way to evaluate individual differences by measuring present and past behavior. For example, the test your instructor will give you to measure your mastery of this material will provide an indication of what you know now. Tests also attempt to predict future behavior. You probably took the Scholastic Aptitude Test (SAT) as part of the admission requirements to college. SAT scores are usually required by higher education institutions as a predictor of success in college.

One important caution needs to be noted here: A test measures only a sample of behavior. Tests are not perfect measures of behavior; they only provide an indication. It is therefore important that case managers not make decisions based solely on test scores.

Types of tests   Thousands of tests are in use today. One way to make sense out of all the tests that are available is to know how they are categorized. One classification is by type of behavior measured. Two categories are identified in this system. *Maximum performance tests* measure ability, and *typical performance tests* give an idea of what an examinee is like. These and other helpful categories are discussed in the pages that follow.

Maximum performance tests include achievement tests, aptitude tests, and intelligence tests. On these tests, examinees are asked to do their best. **Achievement tests** are used to evaluate an individual's present level of functioning or what has previously been learned. Achievement tests that a case manager will often encounter include the Test of Adult Basic Education (TABE) and the Wide Range Achievement Test (WRAT). **Aptitude tests** provide an indication of an individual's potential for learning or acquiring a skill. Because aptitude tests imply prediction, they are useful in selecting people for jobs, scholarships, and admission to schools and colleges. The SAT is an aptitude test. In your work with clients, you will likely read about aptitude tests such as the General Aptitude Test Battery (GATB), the Differential Aptitude Test (DAT), and the Minnesota Clerical Test. When we think about how smart someone is, usually we mean intelligence. Tests such as the Wechsler Adult Intelligence Scale (WAIS), the Wechsler Intelligence Scale for Children (WISC), the Revised Beta Examination (Beta IQ), and the Peabody Picture Vocabulary Test are **intelligence tests**. Careful consideration should be given to these tests and test scores because intelligence can be defined in a number of ways. Some tests measure verbal intelligence, some measure nonverbal intelligence, and others measure problem-solving ability. The WAIS, for example, yields a Verbal IQ, a Performance IQ, and a Full-Scale IQ. On the other hand, the Revised Beta Examination yields only a performance IQ score called a Beta IQ.

The other major category is the typical performance test. Such tests provide some idea of what the examinee is like—his or her typical behavior. In this category are interest inventories (California Picture Interest Inventory, Strong Interest Inventory Test, Kuder Preference Record), personality inventories (Edwards Personal Preference Schedule, Minnesota Multiphasic Personality Inventory, Sixteen Personality Factor Questionnaire), and projective techniques (Rorschach Inkblot Test, Thematic Apperception Test, Rotter Incomplete Sentences Blank). Other well-known performance tests are the Bender–Gestalt Test, the Vineland Social Maturity Scale, and the McDonald Vocational Capacity Scale.

There are a number of other categorization schemes for tests. Individual tests, such as the WAIS or the projective techniques, are administered to one person at a time. Group tests are administered to two or more examinees at a time. The Revised Beta Examination and the Otis Lennon school ability test are group tests, although they can also be administered on an individual basis. Tests can also be classified as standardized or informal. Standardized tests are those that have content, administration and scoring procedures, and norms all set before administration. Informal tests are developed for local use, for example, the test your instructor will give to you to measure your mastery of this course material. Verbal tests use words, whereas nonverbal tests consist of pictures and require no reading skills. Tests in which working quickly plays a part in determining the score are speed tests. Tests such as the Revised Beta Examination are closely timed. In contrast, power tests have no time limit, or one that is so generous that it plays no part in the score.

As you begin to explore the testing literature, you will discover that testing has a language of its own. Recognizing the categories and knowing their meanings will help you develop the vocabulary to understand testing concepts and the advantages and limitations of tests. Selecting tests requires an understanding of the terms in the following list, as well as others.

> *Edition:* the number of times a test has been published or revised
> *Forms:* equivalent versions of a test
> *Level:* the group for which the test is intended (e.g., K–3 is kindergarten through third grade)
> *Measurement error:* the part of an observed test score that is not the true score or the quality you wish to measure
> *Norm:* the average score for some particular group
> *Norms table:* a table with raw scores, corresponding derived scores, and a description of the group on which these scores are based
> *Percentile rank:* the proportion of scores that fall below a particular score

Reliability: the extent to which test scores or measures are consistent or dependable—that is, free of measurement errors

Stanine: a system for assigning the numbers 1 through 9 to a test score

Test: a measurement device

Test administrator: person giving a test

Test profile: a graph that shows test results

Validity: the extent to which a test measures what it claims to measure

Selecting tests   When faced with the task of choosing a test to administer, the first question must be where to find out about available tests. Perhaps the second question, which quickly follows once a case manager realizes the vast number of tests that are available, is how to select a test.

There are many sources of information about tests, ranging from test publishers to reference books available in the library. These sources refer to journal articles that provide more detailed information about a test. The general information about a test can help you narrow the choices to those of interest, and it is for these tests that more specific information is gathered. Let's begin with the more general information.

Thousands of commercially available tests in English are described and critically reviewed in the many editions of the *Mental Measurements Yearbook*, or *MMY*, published by the Buros Institute of Mental Measurements at the University of Nebraska–Lincoln. Begun in 1938 by Oscar K. Buros, the *MMY* provides comprehensive reviews of tests by almost 500 notable psychologists and education specialists (Kaplan & Saccuzzo, 1997). For each test included in the *MMY*, there is a detailed description and price, followed by references to articles and books about the test, along with original reviews prepared by experts. The *MMY* contains no actual tests.

Another reference that summarizes information on tests is *Tests in Print*. This volume is helpful as an index to tests, test reviews, and the literature on specific tests. Entries include the title and acronym of the test, who it was designed for, when it was developed, its subtests, the authors and publishers of the test, and cross-references to *MMY*. Other references may be helpful to you as you narrow your selection, but *MMY* and *Tests in Print* provide the most comprehensive overviews of published tests now available.

Once the choice has narrowed, specimen sets of tests are available for purchase from test publishers. Although there are approximately 400 companies in the test industry, the top 10% are responsible for 90% of the tests used in the United States. Test publishers have catalogs that provide lists of tests and test-related items sold by that company. Companies usually offer specimen sets for sale: the test manual, a copy of the test, answer sheets, profiles, and any other appropriate material related to a particular test. The test manual—the best source of information about a particular test—provides statements about the purposes of the test, a description of the test and its development, standardization procedures, directions for administration and scoring, reliability and validity information, norms, profiles, and a bibliography. This specific information can help the case manager decide whether to use the test.

Once available information about a particular test is gathered, the case manager decides whether to select it. Then the second question, relating to the criteria for selection, surfaces. One helpful source of information is *Standards for Educational and Psychological Testing*, published by the American Psychological Association. The *Standards* is a technical guide that provides the criteria for the evaluation of tests. Among the standards discussed are validity, reliability, test administration, and standards for test use. Any case manager who is involved in testing should carefully review the complete standards.

Criteria for selection   Many considerations go into a decision to use a test. Three primary considerations are validity, reliability, and usability. This section provides a brief overview of each consideration by defining each, identifying the types, and discussing how to evaluate. For more information, consult the *Standards for Educational and Psychological Testing* or testing textbooks that present this material in more detail.

Validity, the most important of the three considerations, is the extent to which a test measures what we actually wish to measure. Think about a time when you have had to move furniture. A yardstick was probably helpful to you as you measured the available space and the piece of furniture for a fit. Long experience has confirmed the validity of a yardstick as a tool for measuring length. Validity works the same way for a test, which is a tool for measuring behavior. Among the questions related to validity are the following: How well do math tests measure math achievement? Could we make predictions based on those scores? Does this test measure other qualities as well? These questions are answered as content-, construct-, and criterion-related validity are established for a test (these terms are likely to appear in the test manual's section on validity).

Validity must be evaluated in light of the intended use. There are a number of factors to consider. One is *evidence* of several types of validity; the more sources of evidence, the better. The quality of the evidence is also important—more important, in fact, than a quantity of questionable evidence. Included in the reported validity data should be a description of the subjects, the testing situation, and how the test was used. Finally, the evidence should relate to the purpose of the test. For example, evidence for the validity of an achievement test should be relevant to content, whereas criterion-related validity is important for tests that predict outcomes.

Reliability, the second consideration, is the degree of consistency with which a test measures whatever it is measuring—the degree to which test scores are free from errors of measurement. Such errors result in inconsistent scores from one test form to another or from one testing time to another, changes that are not attributable to such factors as the examinee's motivation, fatigue, anxiety, or guessing.

As with validity, a review of the reliability evidence of a test takes into account any errors of measurement in light of the test's intended use. For example, split-half reliability, which is often used with longer tests, yields an inflated estimate of reliability when applied to speed tests. Test-retest reliability (which involves testing the same group with the same test on two occasions) would not be appropriate for tests in which memory plays a role from one testing situation to another. When reviewing reliability evidence, it is also important to read the descriptions of the number and characteristics of the sample, the testing situations (including the interval between administrations), and the reported reliability estimates. This information should be evaluated in relation to the test's intended use.

The final consideration is test *usability*, which includes all the practical factors that are part of a decision to use a particular test. The factors include economy, test administration, and interpretation and use of scores. The following questions relate to usability.

ECONOMY
Can we reuse the test booklets?
Are separate answer sheets available?
How easy is it to score the test?

TEST ADMINISTRATION
Are the instructions clear and complete?
Is close timing critical?
What is the layout of test items on the page (print size, clarity of pictures, and so on)?
Can the test be administered via computer?

INTERPRETATION AND USE OF SCORES
Are scoring keys and instructions provided?
Are norms included for appropriate reference groups?
Is there evidence that test scores are related to other variables?
Is there a clear statement about the function of the test and its development?

These questions, as well as the validity and reliability evidence, are all factors to be considered in test selection. Once a test has been selected, the test, answer sheets, scoring keys, interpretive manuals, and so on, are purchased from the test publisher.

Administration and interpretation   The test manual is an important source of information about the administration of a test and the interpretation of test scores. It should identify any special qualifications in training, certification, and experience that are necessary for the people who will administer and interpret the test. Needless to say, only individuals who have the training and experience for using tests should be entrusted with them. The test manual also describes the recording and scoring of answers. Can answers be recorded in the test booklet? Are separate answer sheets available? If the answer to both questions is yes, are the results interchangeable? Also described in the test manual are procedures to facilitate the use of the test and reduce bias. Information about the interpretation of test scores will be included; in some cases, handouts are provided. Note that it is the responsibility of the test administrator to read the manual carefully and be knowledgeable about the test and its uses.

Standardized test administration procedures are important. Many sources of error can influence test scores, and some relate directly to test administration. The testing situation itself is influenced by various factors, including the lighting, ventilation, room temperature, extent of crowding, adequate work space, and noise level. The test administrator's behavior can also be a source of error. Giving extra help, pointing out errors after the test begins, establishing varying degrees of rapport with different examinees, and allowing additional time all influence test scores. The administrator of a test should eliminate as many of these conditions before beginning the test and avoid deviating from the standardized procedures.

Today, computer-assisted test administration is a reality. In this testing situation, test items are presented via a computer terminal or a personal computer, and an automatic recording of test responses occurs. There are two important advantages to testing by computer: (1) standardization of administration and scoring and (2) the control of bias. If a test is administered via computer, it is important to ensure that the items are legible, the screen is free from glare, and the terminals are properly positioned.

What about the client who can't read or who reads below fifth- or sixth-grade level? Or one who has impaired vision or hearing? Or a person with epilepsy who is heavily sedated, with his or her motor processes slowed down? An initial strategy is to find individual psychological tests that can accommodate the client's problems and still measure interests, intelligence, achievement, aptitudes, or personality. The limitations of each disability need to be carefully considered in relation to the person's ability to perform on a test. For example, a client who writes slowly because of a disability should not be given a test that places a premium on speed. In some testing situations, however, special consideration cannot and should not be given to a person with a disability. If a client has an amputation below the knee and is interested in attending college, he or she needs no special considerations when taking the SAT. The limitations imposed by the disability do not interfere with the traits measured by the SAT. When adjustments are made for the person with a disability, avoid deviating from standardized procedures as much as possible. The reason for using tests is to obtain an objective comparison of the individual with a standardized group. If the test is administered in a nonstandardized way, the norm-based scores obtained are not accurate.

When the test is given to people who are different from the groups the test was designed for and normed on, it is more difficult to get accurate results. The instructions, item content and format, methods of answering items, and many other aspects of the test have been designed to make it useful for a specific population. Knowledge of disabilities can help you decide whether modifications are needed. If deviation from established procedures or content is necessary, it is best to consult with a colleague who has psychometric expertise.

The person who takes a test has the right to receive a correct interpretation of the score. It is the responsibility of the test user to interpret scores accurately and meaningfully. Test scores are usually reported as raw scores, which have very little meaning. The raw scores can be converted and reported as standard scores—percentiles, stanines, or T scores—which are easier to understand and interpret. The conversion is based on the norm group or standardization group.

Once the standard scores are available, it is time to interpret the test. There are two essential steps in test interpretation: understanding the results and communicating them to another person, orally or in writing. The following suggestions will guide your preparation for test interpretation.

- Know the test—its purpose, development, content, administration and scoring procedures, validity and reliability, advantages and limitations.
- Avoid technical discussions of tests. Use short, clear explanations of what you are trying to communicate.
- Use the test profile as a graphic presentation of the test results. The examinee may find this easier to follow as the scores are explained.
- Explain what the score means in terms of behavior.
- Go slowly. Give the examinee time to process the information and react.

Tests are helpful tools in measuring traits common to many people. A score serves to show where a person stands in a distribution of scores of peers. How high or low a score is does not measure an individual's worth or value to family, friends, or society. Tyler (1984) suggests a guiding principle for professionals who use tests: The scores are clues to be followed. They do mean something, but in order to know *what*, we must consider each examinee as an individual, combining test evidence with everything else we know about the person. It is unsound practice for case managers to base important decisions on test scores alone. It is important to remember this in test selection, administration, and interpretation.

Summary of testing   Test misuse can easily occur. Let's review some guidelines for the selection, administration, and interpretation of tests.

First, case managers should select tests that they have carefully reviewed. The validity, reliability, and usability of a test; its statement of purpose, content, norm groups, administration, and scoring procedures; and its interpretation guidelines should all be evaluated in light of the intended use. The case manager should check any reviews by experts to add to his or her knowledge of the test. It is also a good idea for the case manager to take the test.

Second, case managers should use only tests they are qualified to administer and interpret. This often depends on one's ability to read the manual. Other tests require advanced coursework and supervision or practicum experiences for proper administration and interpretation. Test catalogs usually indicate how much expertise is required for the tests listed. Another helpful source of information is the *Standards for Education and Psychological Testing*.

Third, case managers who administer tests have an obligation to provide an interpretation of the test results. An understanding of raw scores and their conversion to standard scores, coupled with the ability to communicate the meaning of the scores, is necessary to do this right. In addition, it is essential to be aware of the norm groups and their applicability to the examinee. Some groups, such as

Latinos, African Americans, and rural populations, may be underrepresented in the establishment of norms.

Chapter Summary

This chapter has introduced the planning phase of case management, which includes a number of stages and strategies. Planning begins with a review and continuing assessment of the client's problem. Two areas of concern during this time are a consideration of the problem and the available information about the problem: sources, previous attempts to resolve the problems, and barriers.

The development of a plan guides service provision. Planning is a process of setting goals and objectives and determining intervention. A goal is a brief statement of intent concerning where the client wants to be at the end of the process. Objectives provide the standards by which progress is monitored, the name of the person who is responsible for what actions when, and the criteria for acceptable performance.

Identifying services is a critical part of a well-developed plan. Linking clients and services is facilitated by information and referral systems, which include social service directories, feedback logs, staff training, and updated directories. Some professionals prefer setting up their own information and referral systems.

Part of the planning process or the development of the service plan may be gathering additional information. Two data collection methods used by case managers are interviewing and testing. Assessment interviews, structured clinical interviews, case history interviews, and mental status examinations are the four types of interviews. Potential sources of error include the halo effect and general standoutishness.

Testing, a second data collection method, is encountered by the case manager in reports, in case files, and as a service offered by an agency. Understanding test language, concepts, sources of test information, and the factors included in selecting, administering, and interpreting tests facilitates the meaningful use of test scores.

Chapter Review

◆ *Key Terms* ....

Plan	*Case history interview*
Goals	*Sources of error*
Objectives	*Halo effect*
Plan development	*Test*
Information and referral system	*Psychological test*
Social service directory	*Maximum performance test*
Feedback logs	*Achievement test*
Assessment interview	*Aptitude test*
Structured clinical interview	*Intelligence test*

◆ Reviewing the Chapter ....

1. Describe the two areas of concern addressed by revisiting the assessment phase.
2. What sources help a case manager review a client problem?
3. What role does documentation play in the review of the problem?
4. Define *plan*.
5. What activities occur before development of the plan?
6. List the characteristics of a service plan.
7. What are the benefits of establishing goals?
8. List the criteria for well-stated and reasonable goals.
9. Distinguish between a goal and an objective.
10. Identify a problem you would like to address, and develop a plan with goals and objectives.
11. List the three components of information and referral and give an example of each.
12. Discuss the similarities between interviewing and testing.
13. Compare the four types of interviews and their roles in case management.
14. Illustrate how sources of error may affect an interview.
15. How do case managers use tests?
16. Describe the different ways to categorize a test.
17. Describe how to select a test to measure vocational interests.
18. Define *validity* and *reliability*, and describe their roles in test selection.
19. What makes the test manual the best source of information about a test?
20. Identify some sources of error in testing.
21. Under what conditions should special considerations be made for a person with a disability?
22. Identify the two essential steps in test interpretation.

◆ Questions for Discussion ....

1. Why do you think developing a plan is important?
2. If you were a new case manager, how would you begin to develop a network of available services?
3. What kinds of criteria would you use to determine whether you need to conduct a structured interview with an 8-year-old?
4. Do you believe that you will be able to determine what errors exist in the information that you gather? What problems do you expect to encounter in finding errors?

References

Boserup, D. G., & Gouge, G. (1977). *The case management model*. Athens, GA: Regional Institute of Social Welfare.

Hollingsworth, H. L. (1922). *Judging human character*. New York: Appleton-Century-Crofts.

Kaplan, R. M., & Saccuzzo, D. P. (1997). *Psychological testing: Principles, applications, issues.* Pacific Grove, CA: Brooks/Cole.

McClam, T. (1992, September–October). Employer feedback: Input for curriculum development. *Assessment Update,* 4(5), 9–10.

McClam, T., & Woodside, M. R. (1994). The practitioner's voice: Case management for effective service delivery. *Human Service Education,* 14(1), 39–45.

Thorndike, E. L. (1920). A constant error in psychological rating. *Journal of Applied Psychology, 4,* 25–29.

Tyler, L. E. (1984). What tests don't measure. *Journal of Counseling and Development, 63,* 48–50.

Chapter Seven ....

Building a Case File

*A*nd in this particular job, it is useful to have working knowledge of medical terms, to know how to use a Physician's Desk Reference and a pediatric manual for definitions, and to have a basic understanding of things like physical, occupational, and speech therapies.
> —Margaret Mikol, SKIP of New York City, personal communication, May 4, 1994

*W*e are a short-term shelter with a maximum of 30 days. We try to give them as much as possible during that time. The residents are involved in an assessment completed by the psychologist; they participate in psycho-educational groups and receive individual and family therapy.
> —Stacie Newberry, Marion Hall Emergency Shelter, St. Louis, MO, personal communication, December 13, 1999

*W*hen a referral comes in, the first thing we do is contact the school and say give me a profile of this student. So they come with a problem . . . same thing with the medical . . . we get their student profile and look at how they are doing academically and emotionally.
> —Norma Cervantes, Roosevelt High School, Los Angeles, personal communication, March 24, 1998

Information from other professionals comes to the case manager in two ways. When he or she receives a case file on a client from another agency or worker, it may contain reports or evaluations from other professionals. In other situations, the plan developed by the case manager and the client may include referrals to other professionals for evaluations. In both situations, the case manager must be able to understand the information provided and (if asking for help from other professionals) to know just what to request.

The chapter-opening quotations illustrate the kinds of information that may be needed from other professionals in order to develop a plan or to provide services. The medical information, histories, or exams these three helpers mention are part of the case files of clients who have medical problems. Margaret Mikol speaks of the advantages of being familiar with medical terms and medical references when trying to decipher medical reports. Physical assessments and psychological assessments offer important information to the Marion Hall staff as they work with homeless and runaway female teens.

Professional staff at Roosevelt High School in Los Angeles gather as much information about the student from other schools.

This chapter examines the types of information that may be found in a case file or that must be gathered to complete one. Exactly which information is needed depends on the individual's case and the agency's goals, but many cases involve medical, psychological, social, educational, and vocational information. We introduce each type of information, give a rationale for gathering it, describe the kinds of data likely to be provided, and discuss what the case manager needs to know in order to make the best use of the report. For each section of the chapter, you should be able to accomplish the following objectives.

MEDICAL INFORMATION
- Tell how medical information contributes to a case.
- Decode medical terms.

PSYCHOLOGICAL EVALUATION
- List the reasons for a psychological evaluation.
- Make an appropriate referral.
- Identify the components of a psychological report.
- Describe the type of information provided by the DSM-IV.

SOCIAL HISTORY
- State the advantages and limitations of a social history.
- Name the topics included in a social history.
- List the ways social information may appear in the case file.

OTHER TYPES OF INFORMATION
- List the types of educational information that may be gathered.
- Define a vocational evaluation.

Medical Information

Knowledge of medical terminology, conditions, treatments, and limitations is important in understanding a case. Medical information may be provided on a form or in a written report. The exam and report may have been done by a general practitioner or by a specialist in a field such as neurology, orthopedics, or ophthalmology. In some cases, the case manager can interact with the medical service provider and thus be able to ask questions, request specific assistance, or offer observations. Often, however, he or she does not have this opportunity and must rely on the written report. Then the resources mentioned at the beginning of the chapter may prove particularly helpful. Many agencies have a copy of the *Physician's Desk Reference* (PDR) or other medical guide. Some also have a physician serving as a consultant, who is available to answer questions. This section introduces basic medical information to help you understand medical terminology.

Agencies approach medical information in different ways. Some require documentation of a mental or physical disability or condition in determining eligibility for services. Others use a medical examination as part of their assessment procedures. In certain situations, medical information is not gathered unless there is some indication or symptom of a disease, condition, or poor health that would affect service delivery.

Medical knowledge is particularly crucial when working with people who have disabilities. A general medical examination and specialists' reports help determine the person's functional limitations and potential for rehabilitation. It is important to set objectives that are realistic in light of the client's physical, intellectual, and emotional capacities. When a medical report covers a disability in functional terms, "it addresses the following factors: the description can read like the following: strength, climbing, balancing, stooping, kneeling, crouching, crawling, reaching, handling, fingering, feeling, talking, hearing, tasting, and smelling, near acuity, far acuity, depth perception, visual accommodation, color vision, and field of vision" (Dabatos, Rondinelli, & Cook, 2000, p. 81).

Each medical evaluation includes recommendations including the individual's physical, emotional, and intellectual capacities. Following is a sample medical recommendation.

The individual has a diagnosis of obsessive-compulsive disorder and has limited strength, balancing, hearing, and near-acuity functionality. This person needs work with supervision, few stressors, limited lifting, and limited need for close work.

Often, however, the form for a general medical examination allows only a small space for the diagnosis, so the case manager reads a phrase such as "chronic back pain," "normal exam," or "emotional problems." Not very helpful, is it? Remember that the client is an important source of information; he or she can tell you about any problems. You may then need to decide whether or not a specialist's evaluation would be helpful.

Medical Exams

Generally, medical information contributes to a case in two ways. **Medical diagnosis** appraises the general health status of the individual and establishes whether a physical or mental impairment is present. For example, 10-year-old Javier Muldowny comes into state custody, abandoned by his parents. The case manager at the Assessment, Care, and Coordination Team takes Javier to the Health Department for an examination. The examination results in a diagnosis of otitis media.

Diagnostic medical services include general medical examinations, psychiatric evaluations, dental examinations, examinations by medical specialists, and laboratory tests. A medical diagnosis is helpful when the client has a medical problem or is currently receiving treatment from a physician, who may provide important information about social and psychological aspects of the case in addition to the medical aspects. When making a referral for a medical diagnosis, the case manager should help the client understand why the referral is necessary, the amount of time it will require, what the client can expect to learn, and what use the agency will make of the report.

Medical consultation is used in several ways. First, the consulting physician can provide an interpretation of medical terms and information. For example, Javier Muldowny was diagnosed with otitis media. The case manager received this report, asked a colleague what the diagnosis meant, and learned that it was an ear infection. A consultation with a physician would reveal that otitis media is a severe ear infection that sometimes results when the Eustachian tubes are not properly angled. The consultation might also explain the report further and clarify possible treatments. In Javier's case, the case manager may need further information about the advantages and disadvantages of two possible treatments: insertion of tubes in the ears and a regimen of antibiotics. A consultation with an otorhinolaryngologist (ear, nose, and throat specialist) could shed light on the medical prognosis and the extent of any hearing disability that might be expected.

The role of a medical consultant is to interpret the available medical data, determine any implications for health and employment, and recommend further medical care if needed. The case manager can make the best use of a consultant by being prepared for the meeting, perhaps specifying in writing what is needed from the consultant. This usually involves identifying problems that need to be resolved and setting forth the significant facts of the case. The case manager needs to understand medical terminology, the skills of specialists in diagnostic study and treatment programs, and the effects of disability on a client.

The medical service used most often in human services is the **physical examination**, in which a physician obtains information concerning a client's medical history and states the findings. The exam data are entered into the medical record. Here we give an overview of the physical examination: the kinds of information obtained and what the case manager needs to know to make such a referral and to understand the physician's report.

Diagnosis involves obtaining a complete medical history and conducting a comprehensive physical exam (also called a physical, a health exam, or a medical exam). The results of the exam may be reported on a form provided by the referral source. Sometimes physicians use preprinted schematic drawings of various body parts or organ systems to enhance or clarify the written report. However the information is transmitted, the quality of the reporting depends on the relationship between the physician and the patient. In some cases, the patient has mixed feelings about the referral for a physical exam. He or she needs an explanation of why the referral is necessary, the amount of time the exam will take, what outcome is expected, and how the information will be used. Keep in mind that the client's socioeconomic status, language skill limitations, or cultural background may also influence how he or she feels about the referral. If it is communicated with sensitivity, and if a good relationship with the physician is established, any barriers of anxiety, depression, fear, or guilt can be overcome.

The general medical exam is done by a physician who takes an overall look at the person's medical state. Its purpose is to evaluate the person's current state of health, focusing on two areas. First, a complete medical history records all the factual material, including what the client states and the physician's inferences from what is not said. A typical starting point is the chief complaint, as expressed by the individual. (See Figure 7.1.) If there is an illness at present, it is described in terms of onset and symptoms (including location, duration, and intensity). A family history relates significant medical events in the lives of relatives, particularly parents, grandparents, siblings, spouse, and children. Extensive information about the individual's past medical history is also collected. This may include childhood diseases, serious adult illnesses, injuries, and surgeries. A review of symptoms focuses on information about present and past disorders, which the physician elicits through questions about organs and body systems. After completing the physical exam, the physician records a diagnostic impression. The actual diagnosis is made once there is conclusive evidence, which may mean getting further studies or referring the client to a specialist for consultation.

What exactly makes up a medical exam? Techniques used during a physical exam are inspection, palpation (feeling), percussion (sounding out), and auscultation (listening). Usually, the examining physician works from the skin inward to the body, through various orifices, and from the top of the head to the toes (Felton, 1992). Special instruments are used to look, feel, and listen. More time is spent in particular areas to ascertain whether a certain finding truly represents a change in an organ or tissue. Some parts of the exam are carried out quickly, and others require more time. More important areas may receive a second, more thorough examination. The physician records the findings as soon as possible after completing the exam and shares the results with the client.

For some clients, one of the first things that occurs in the case management process is a referral to a physician for a general medical exam. As the physician conducts the exam, he or she completes a form like the one shown in Figure 7.2, which is then sent to the referring counselor. It becomes part of the client record.

Medical Terminology

Medical reports often include **medical terminology**, which may seem like a foreign language to a case manager who is unfamiliar with it, because physicians rely on technical words and phrases for exactness. Medical specialties also have special terminologies. Other professionals who may write reports using medical terminology are nurses, physical therapists, and occupational therapists. It can be a challenge for the case manager to make sense of these reports; he or she must have at least a rudimentary understanding of medical terminology.

Medical terminology follows simple rules. To analyze medical words, identify the four elements that are used to form such words: the word root, the combining form, the suffix, and the prefix. It may help to think of these elements as verbal building blocks. Let's examine each component.

MEDICAL REPORT

CHIEF COMPLAINT: Weakness and malaise

PRESENT ILLNESS: Three weeks ago, this 40-year-old single African American male had a cold, with associated mild cough and temperature elevation, which lasted two days. At that time there was a loss of appetite and a decrease in food intake.

After the cough and temperature elevation subsided, the patient noted increased weakness and general malaise. The patient reports he tires easily and is unable to sustain exercise, which was tolerated well before the onset of the cold.

FAMILY HISTORY: This patient's mother, 75, has had breast cancer and a radical mastectomy and was diagnosed with lupus ten years ago. His father, 80, had quadruple bypass surgery five years ago. There is extensive evidence of heart disease in father's family. There is a history of diabetes and tuberculosis in mother's family, although the mother has had neither.

PAST HISTORY: Patient had chickenpox at age 7. Sustained multiple fractures in a motor vehicle accident at age 16, but he is without permanent motor or neurological damage. Frequent lower respiratory infections characterized by productive cough and yellow-mucus-producing hacking cough. The patient has been exposed to tuberculosis, showing a positive PPD test but negative chest X ray five years ago, and was treated with usual course of preventative medications. Since that time patient has been free of other persisting symptoms.

SOCIAL HISTORY: Unmarried, owns home, smokes one and a half packs of cigarettes a day, drinks occasionally.

OCCUPATIONAL HISTORY: Worked construction jobs following high school. Has experience with foundation work, masonry, and plumbing. Last spring he graduated from a local college with degree in accounting. Plans to start his own construction business next year.

Figure 7.1   Medical report

REVIEW OF SYSTEMS:

HEAD, EYES, EARS, NOSE, THROAT: No frequent or severe headaches or head injuries. Wears corrective lenses for nearsightedness. Had frequent ear infections as child, but has suffered no hearing loss. Has two or three colds a year, but is without sinus pain. No problems with throat.

NECK: No significant abnormality.

RESPIRATORY TRACT: No expectoration of blood. Wheezing and shortness of breath with exertion—walking up stairs. When colds "go to chest," takes over-the-counter cough medications. Early morning nonproductive cough present. States he is not trying to quit smoking.

CARDIOVASCULAR SYSTEM: No chest pain or palpitations. No history of murmur, coronary artery disease, or hypertension. Fatigue has been increasing over the last month, making him unable to complete his morning walk, which is normally 2 miles.

GASTROINTESTINAL SYSTEM: Appetite has decreased since this cold. No indigestion. States he is on no special diet.

GENITOURINARY SYSTEM: Denies venereal disease.

NERVOUS SYSTEM: No significant abnormality.

MUSCULOSKELETAL SYSTEM: Negative.

ENDOCRINE SYSTEM: Negative.

Figure 7.1   *(Continued)*

GENERAL BASIC MEDICAL EXAMINATION RECORD

This record is CONFIDENTIAL

Section I. - (To be filled out by rehabilitation agency)

Client No. _____

Johnson, _Roy_ _R._   _7/16/60_   _W_   _m_   S _✓_ M___ D___ Sep_
(Last Name)   (First Name)   (Middle Name)   (Date of birth)   (Race) (Sex)   (Martial Status)

Rt. 1   Box 68   _Centerville_   _TN_
(Home address: Street and number or R.F.D.)   (City or town)   (County)   (State)

Usual occupation _Plumber_   Description of last job _Same_

Last time hospitalized _5-XX_   _Surgery_   _____
(Date)   (Reason)   (Name and location of hospital)

Last visit to physician _18 mos._ _Spinal cord injury_   _Dr. Alderman_
(Date)   (Reason)   (Name and address of physician)

Is patient now under care of physician? _yes_   _Dr. Brown_
(Yes or no)   (If answer is "Yes," give name and address of physician)

Patient's statement of disabilities _Spinal cord injury_

Signature of rehabilitation counselor _Tom Chapman_   Date _7-20-XX_

Section II. PERTINENT HISTORY (To be filled out by physician.)

Back surgery; surgery twice
continued pain

Section III. PHYSICIAN EXAMINATION. (To be filled out by physician. Items checked ✓ were examined and found normal. Deviations from normal are noted. If items require additional description, please record on extra sheet.)

Height (without shoes): _5_ ft. _10_ in.   □ Weight (without clothing) _220_ pounds   □ Temperature _98⁴_

Eyes - Right _____✓_____ LEFT_____✓_____
(Discharge; corneal scars; strabismus; pterygium; ptosis; trachoma; fundi; cataract; intraocular tension)

Distant vision: Without glasses: R. 20/_30_ L. 20/_70_. With glasses: R. 20/_20_ L. 20/_20_
(If vision is too low to be recorded at 20 feet, indicate by recording as "less than 20/200")

Ears - Hearing: Right _20_ Left _20_   □ Other findings: R_____✓_____ L_____✓_____
20 feet   20 feet   (Evidence of middle ear or mastoid disease. Drums: Normal, absent
(Consider denominators here indicated as normal   perforated, dull, retracted, discharge).
Record as numerators greatest distance heard)

Nose _____✓_____   □ Throat _____✓_____
(Obstruction, evidence of chronic sinus infection, polypi   (Tonsils: Normal, enlarged, removed, etc.)
perforated septum, etc.)

Mouth_____✓_____   □ Neck _____✓_____
(Missing teeth, pyorrhea; abnormality of tongue or palate)   (Thyroid enlargement, nodules, etc.)

Lymphatic system____✓_____   □ Breasts _____✓_____
(Especially cervical, epitrochlear, inguinal)   (Abnormal discharge, nodules, tenderness, hypoplasia)

Lungs: Right _____✓_____   □ Left _____✓_____
(If history or physical findings indicate active or arrested tuberculosis, recommend chest X-ray, sputum examination, and consultation with chest specialist)

Figure 7.2   Medical examination form

Circulatory System: Heart _____ ✓ _____
(Enlargement, thrill, murmurs, rhythm)

Blood pressure { Systolic } 118/80 _____ Pulse rate 76 _____ Dyspnoea O _____ Cyanosis O _____ Edema O _____
{ Diastolic }

Evidence of arteriosclerosis _____ None _____
(Type; degree; where found, as "cerebral," "brachial," etc.)

Abdomen _____
(Scars, masses, palpable liver, palpable spleen, etc.)

Hernia _____ ✓ _____
(Type: Inguinal, ventral, femoral, etc. Right, left, bilateral)

Genito-urinary _____ ✓ _____
(Urethral discharge, varicocele, hydrocele, scars, epididymitis, enlarged or atrophic testicle)

and
Gynecological _____
(Prolapse, cystocele, rectocele, Cervix)

Ano-rectal _____ ✓ _____
(Hemorrhoids, prolapse, fissures, fistula. Prostate)

Nervous system _____ ✓ _____
(Paralysis, Sensation, Speech, Gait, Reflexes: Pupillary, Knee, Babinski, Romberg)

_____ ✓ _____ (Memory, Peculiar ideas or behavior. Spirits: Elated, depressed, normal)

_____ ✓ _____
(Neurological or psychiatric abnormalities should be described on separate sheet)

Skin _____ ✓ _____   □ Feet _____ ✓ _____   □ Varicose veins _____ ✓ _____
(Moist, dry, clear)   (Weak feet, Congenial or traumatic defects)   (Site)

Orthopedic Impairments: (*Describe*) _____ Low back pain _____

Laboratory: Urinalysis: Date 8-7-XX _____ □ Specific gravity 1.020 _____ □ Reaction 7 _____
□ Albumen neg _____ □ Sugar neg _____

DIAGNOSIS: (Indicate major and minor) _____ Chronic lower back pain after two spinal _____
_____ ops for herniated disc _____

STATUS: of major disability: (Check appropriate terms) Permanent ✓ _____ Temporary _____ Stable _____

Slowly progressive _____ Rapidly progressive _____ Improving _____

PROGNOSIS: Can the major disability be removed by treatment? □ ☑ Substantially reduced by treatment? ☑ □
                                                            (Yes) (No)   (Yes) (No)

Physical Capacities: (Under "Physical activities" and "Working conditions" use symbols as follows:
(✓) No limitation. (X) Limitation. (0) To be avoided.

Physical activities: Walking X _____ Standing X _____ Stooping X _____ Kneeling X _____ Lifting X _____ Reaching X _____ Pushing X
Pulling X _____ Other (*Specify*) _____

Working conditions: Outside ✓ _____ Inside ✓ _____ Humid ✓ _____ Dry ✓ _____ Duty ✓ _____ Sudden temperature changes ✓
Other (*Specify*) _____

RECOMMENDATIONS:

□ Is examination by specialist advisable? If so, specify which speciality _____

□ Refraction   □ X-ray of chest   □ Other diagnostic procedures (*Specify*) _____
□ Prosthetic appliances (*Specify*) _____
□ Hospitalization (*Specify reasons and approximately duration*) _____
□ Treatment (*Specify type and approximately duration*) _____

Remarks: Please use additional sheet for remarks and expansion of any observations.

Date 8-7-XX _____ _____ John H Jones _____ M.D.
                                                              (Physician)

_____ Suite 201 Physicians Office Bldg. _____
                                              (Address)

Figure 7.2   (*Continued*)

Word roots   The main part or stem of a word is the **word root**. In medical terminology, the root usually derives from Greek or Latin and often indicates a body part. All medical words have one or more word roots.

GREEK WORD	MEANING	WORD ROOT
kardia	heart	cardi
gastro	stomach	gastr
nephros	kidney	nephr
osteon	bone	oste

Combining forms   A word root plus a vowel, usually an *o*, is the combining form, as in the following examples.

WORD ROOT		COMBINING VOWEL		COMBINING FORM	MEANING
cardi	+	o	=	cardio	heart
gastr	+	o	=	gastro	stomach
nephr	+	o	=	nephro	kidney
oste	+	o	=	osteo	bone

Suffixes   A *suffix* is a word ending. In medical terminology, the suffix usually denotes a procedure, condition, or disease, as in the instances listed here.

COMBINING FORM		SUFFIX		MEDICAL WORD	MEANING
arthr (joint)	+	-centesis (puncture)	=	arthrocentesis	puncture of a joint
thoraco (chest)	+	-tomy (incision)	=	thoracotomy	incision in the chest
gastro (stomach)	+	-megaly (enlargement)	=	gastromegaly	enlargement of the stomach

Suffixes also form adjectives, express relative size, indicate surgical procedures, and express conditions or changes related to pathological processes. Examples follow.

FORM ADJECTIVES	EXAMPLE	MEANING
-al (means "pertaining to")	arterial	pertaining to an artery
-ible (indicates ability)	digestible	capable of being digested

EXPRESS RELATIVE SIZE	EXAMPLE	MEANING
-ole (means small)	arteriole	a small artery
-ule (means small)	granule	a small grain

INDICATE A SURGICAL PROCEDURE	EXAMPLE
-ectomy (means "removal of an organ or part")	appendectomy

EXPRESS CONDITIONS OR CHANGES RELATED
TO PATHOLOGICAL PROCESSES
-mania (means "excessive excitement or
  obsessive preoccupation")

EXAMPLE
pyromania

Prefixes   The word element located at the beginning of a word is the *prefix*. It usually denotes number, time, position, direction, or negation.

PREFIX		WORD ROOT		SUFFIX		MEDICAL WORD	MEANING
hyper (exces- sive)	+	therm (heat)	+	ia (condi- tion)	=	hyper- thermia	(condition of excessive heat)
micro (small)	+	card (heart)	+	ia (condi- tion)	=	microcardia	(condition of a small heart)

Other common prefixes that modify word roots indicate position (e.g., *ab* means "away from," as in *abnormal*), quantitative information (e.g., *a* or *an* means "without," as in *anorexia*, or without appetite), qualitative information (e.g., *mal* means "bad," as in *malfunction*), and sameness or difference (e.g., *homo* or *hetero*). For other prefixes and suffixes that are common in medical terms, see Table 7.1. There are three basic steps to working out the meaning of a medical term. First, identify the suffix and its meaning. Second, find the prefix, if any, and determine what it means. Third, identify the root words and their meanings. For example, *thermometer* consists of a suffix (*meter*, meaning "instrument for measuring") and a word root (*thermo*, meaning "heat"). Thus, a thermometer is an instrument for measuring heat. Another example is *gastroenteritis*. The suffix is *itis* (inflammation), the prefix is *gastr* (stomach), and the word root is *enter* (intestine). Gastroenteritis is an inflammation of the stomach and intestine. Remember that the vowel *o* is a combining form, linking one word root to another to form a compound word. *Osteoarthritis* is another example. The suffix *itis* means "inflammation"; word roots are *oste*, which means "bone," and *arthr*, which means "joint."

TABLE 7.1   COMMON PREFIXES AND SUFFIXES

Prefix	Meaning	Suffix	Meaning
dys-	bad, painful, difficult	-itis	inflammation
macro-	large	-algia	pain
hypo-	under, below	-toxin	poison
scler-	hard	-oma	tumor
tachy-	rapid	-pathy	disease
hyper-	over, above, excessive	-osis	abnormal condition, increase

TABLE 7.2   SOME COMMON COMPONENTS OF MEDICAL TERMS

Component	Meaning	Example
-algia	pain	euralgia
angio-	blood vessel	angiogram
arth-	joint	arthroscopy
contra-	opposed to	contraception
derm-	skin	dermatology
-emia	condition of the blood	polycythemia
enceph-	brain	encephalitis
glyco-	sugar	glycosuria
hepat-	liver	hepatitis
hyster-	uterus	hysterectomy
leuk-	white	leukocyte
lip-	fatty	perlipemia
-oscopy	visual examination	laparoscopy
-ostomy	creation of an artificial opening	tracheostomy
-otomy	incision	craniotomy
-plasty	reparative or reconstructive surgery	rhinoplasty
pre-	before	precancerous
pyel-	pelvis	pyelogram
syn-	together	synarthrosis
tri-	three	triceps

The *o* is the combining vowel. Osteoarthritis means inflammation of bone and joint. The following list contains some common medical terms that use suffixes, prefixes, and word roots introduced in this chapter. Can you fill in the columns to work out the meaning of each term? Other examples are shown in Table 7.2.

TERM	SUFFIX/PREFIX	WORD ROOT	MEANING
tachycardia			
dysfunction			
gastritis			
nephritis			
osteopathy			
hypodermic			

It is a continuing challenge for case managers to keep current with terminology because of ambiguities, inconsistencies, and the changing course of medical knowledge. Although most word roots have Greek or Latin origins, some occur in both but have different meanings. The root *ped*, for example, means "child" in Greek (e.g., pediatrician), but in Latin *ped* means "foot" (e.g., pedicure). Many diseases are named for individuals, such as Alzheimer's disease and Hodgkin's disease. Some disorders are called syndromes: Cushing's syndrome, Horner's syn-

TABLE 7.3   MEDICAL ABBREVIATIONS

Abbreviation	Meaning	Abbreviation	Meaning
a.c.	before meals	L-1, L-2, L-3	lumbar vertebrae (by number)
b.i.d.	twice daily	LLQ	left lower quadrant
B.P.	blood pressure	LMP	last menstrual period
C-1, C-2, C-3	cervical vertebrae (by number)	p.c.	after meals
CBC	complete blood count	p.r.n.	as needed
CNS.	central nervous system	q.i.d.	four times daily
DX	diagnosis	RLQ	right lower quadrant
F.H.	family history	RX	treatment
GI	gastrointestinal	S-1, S-2, S-3	sacral vertebrae (by number)
GU	genitourinary	T-1, T-2, T-3	thoracic vertebrae (by number)
HDL	high-density lipoprotein	t.i.d.	three times daily
h.s.	at bedtime	WBC	white blood count
H & P	history and physical examination		

drome. Acronyms are formed from the initials of lengthy terms: MRI (magnetic resonance imaging) and ACTH (adrenocorticotropic hormone) are examples. In addition, medical terminology traditionally uses hundreds of abbreviations; some of the most common are listed in Table 7.3. Keeping informed about trends in medicine increases one's understanding of the meanings of terms. For example, physicians increasingly prescribe generic drugs rather than brand names (e.g., the generic diazepam rather than Valium). Keeping current with medical terminology entails awareness of chemicals, syndromes, and diseases that are newly named and sometimes given acronyms or abbreviations (e.g., AIDS for acquired immunodeficiency syndrome). It must also be remembered that words can have multiple meanings and that several names may apply to a single entity.

 ## Psychological Evaluation

The objective of a **psychological evaluation** is to contribute to the understanding of the individual who is the subject. The report writer is a consultant who makes a psychological assessment that is practical, focused, and directed toward the solution of a problem. The psychological report he or she prepares is more than a presentation of data. This section helps you determine when a psychological evaluation is needed, how to make the referral, and how to prepare the client. The evaluation itself and the report are also discussed.

Referral

Case managers may refer clients for psychological evaluations for a number of reasons. One reason is to establish a diagnosis in order to meet criteria of eligibility for services.

Nadine is a deeply depressed 15-year-old who is currently taking antidepressant medication. She is increasingly out of control. Yesterday, she slapped her grandmother, with whom she lives, and threatened to kill her. If she is to receive services in an inpatient treatment program, she must have a diagnosis confirming emotional disturbance.

Another reason for a psychological evaluation is to provide justification for a particular service.

Amal is a 28-year-old male whose divorce will be final in a month. As the court date approaches, Amal feels more and more depressed. He is having trouble getting up in the morning, showing up for work on time, and maintaining relationships with those who are close to him. His physician has suggested counseling, but Amal's insurance company insists that he have a psychological evaluation to determine whether or not he needs it.

Sometimes a psychological evaluation functions as a screening or routine evaluation to obtain information about a client's personality, aptitude, interests, intelligence, and achievement.

Greg is a 35-year-old male who is the only child of elderly parents. He is mentally retarded. His parents, concerned about who will care for Greg if something happens to them, have learned of a group home where the residents live under close supervision. One requirement for acceptance into the program is a recent psychological evaluation that assesses intelligence as well as ability to function independently.

A case manager may also order a psychological evaluation to resolve contradictions or ambiguities or to add information that is missing.

Paloma is a 10-year-old who is enrolled in public school. Her teacher is concerned about her behavior. One day she is passive, rarely interacts with her classmates, and does not participate in class. The next day, she may be loud, talkative, and disruptive. Just yesterday, she started a fight with a classmate. This has prompted her teacher to request an evaluation from the school psychologist.

Finally, a psychological evaluation may be recommended to answer particular questions regarding the client.

Is there brain damage? Why does the individual have trouble relating to others? How is this person adjusting to the recent amputation of her leg? Why is the client doing poorly in school?

In any of these situations, a referral for a psychological evaluation is appropriate. In each case, the case manager seeks help in order to provide the client with needed services. It is easiest to get what is needed if the consulting psychologist knows the general mission of the agency and understands the specific problem to be addressed. Having this information allows him or her to choose the most relevant and efficient approach to gathering the needed information. The referral for a psychological evaluation is usually made by a case manager, who specifies what is needed: a routine workup, testing, questions about the case, a diagnosis. Thus, the psychologist is charged with a mission. It is therefore critical that the referral be more than a general request, such as "psychological evaluation" or "for psychological testing." These terms communicate poorly; the referring professional has failed to express what prompted the referral. Two scenarios may result: The psychologist may ask the case manager for more specific information, or he or she may try to guess what is wanted or needed. When the reason for the referral is not clear, it is difficult for the psychologist to provide a useful report.

How does a case manager make a good psychological referral? First, it is important to be clear about the reason for referral. The case manager must clarify the need for documentation of a condition or disability, obtaining test scores, or the exploration of behavioral inconsistencies. Specific questions also help the psychologist focus on the client's problems. The psychologist then makes recommendations to the case manager. The two professionals can discuss the case before the evaluation to clear up any questions or needs. Since many referrals are made by phone or direct personal contact, such a discussion can easily take place, but it may be even more important when the referral is made in writing.

Part of making a successful referral is preparing the client for the psychological evaluation. To do this, the case manager needs a clear understanding of the process and the ability to explain it to the client. Some clients may be suspicious of testing or may fear that the case manager considers them crazy. Demystifying the evaluation helps to dispel these attitudes.

The Process of Psychological Evaluation

The evaluation itself includes a study of past behavior, conclusions drawn from observations of current behavior, a diagnosis, and recommendations. This study requires the psychologist to assess which data are important to the client's presenting problems. In some cases, relevant information is in the client file; it is then helpful for the psychologist to have access to these documents in addition to the observations and questions from the referral source.

One of the primary ways that a psychologist observes current behavior is by testing. From the discussion of testing in the previous chapter, you know that testing gives samples of behavior. That discussion also introduced a number of tests that are useful in human services. Psychologists use many of them, notably the WAIS and projective tests (such as the Rorschach and Thematic Apperception Test). These tests are individually administered and scored, and psychologists are specially trained to use them. As a consultant, then, the psychologist decides what kinds of data must be gathered to carry out the assignment given by the referral source, which findings have relevance, and how these findings can be most effectively presented.

The results of the psychological evaluation are communicated to the case manager in a written report. The **psychological report** is "a document written as a means of understanding certain features about a person and his or her current life circumstances in order to make decisions and to intervene positively in a problem situation" (Tallent, 1993, p. 27). The report may appear in one of several forms, the most common of which is a narrative (illustrated by the report included in this section).

Results may also be communicated as a terse listing of problems and proposed solutions. Another option is the computer-generated report, usually consisting of a sequence of statements or a profile of characteristics. Less frequently used are checklists of statements or adjectives, clinical notes, and oral reports relating impressions. Since the narrative is the form of psychological report that is most often used in human services, let's explore it further.

Usually, the content, sources, and format of narrative psychological reports follow a similar pattern. There are three components to the content of a report. One is the orienting data, which includes the reason for the referral and pertinent background information, such as age, marital status, social history, and educational record. Illustrative and analytical content is the second component; here one finds the interpretation of raw data, including test scores. The third component, the psychologist's conclusions, includes a diagnosis and recommendations, which are presented with supporting evidence. The sources of the information in all three components are the interview between the psychologist and the client; test data; behavior observed during the evaluation; any available medical reports and social histories; and any observations, case notes, or summaries written by other professionals involved with the case.

Among the headings that organize the report are Reason for the Referral, Identifying Data, and Clinical Behavior. Under such headings one would find the reason for the assessment, identifying information, any social data, and the psychologist's observations of behavior during the evaluation. The subsequent headings—Test Results, Findings, Test Interpretation, or Evaluation—may be subdivided into Intellectual Aspects (e.g., an IQ score and what it means) and Personality (e.g., psychopathology, attitudes, conflicts, anxiety, and significant relationships). The Diagnosis section presents the main evaluative conclusions, usually expressed as a series of numbers followed by the name of a disorder or condition. The classification system for diagnoses used in the United States is

TABLE 7.4   EXCERPT FROM THE DSM-IV: CODES FOR ADJUSTMENT DISORDERS AND ATTENTION-DEFICIT DISORDER/HYPERACTIVITY DISORDER

Adjustment Disorder

309.0	With Depressed Mood
309.24	With Anxiety
309.28	With Mixed Anxiety and Depressed Mood
309.3	With Disturbance of Conduct
309.4	With Mixed Disturbance of Emotions and Conduct
309.9	Unspecified

Attention-Deficit Disorder/Hyperactivity Disorder

314.01	Attention-Deficit/Hyperactivity Disorder, Combined Type
314.00	Attention-Deficit/Hyperactivity Disorder, Predominantly Inattentive Type
314.01	Attention-Deficit/Hyperactivity Disorder, Hyperactive-Impulsive Type

SOURCE: Reprinted with permission from *Diagnostic and Statistical Manual of Mental Disorders*, Fourth Edition. Copyright 1994 American Psychiatric Association.

published by the American Psychological Association in the *Diagnostic and Statistical Manual of Mental Disorders*, Fourth Edition (DSM-IV).

The **DSM-IV** codes include a broad range of psychological disorder categories, such as adjustment disorders, alcohol, attention-deficit/hyperactivity disorder, cocaine, major depressive disorder, and schizophrenia. Table 7-4 presents a detailed list of adjustment disorders and attention-deficit/hyperactivity disorders taken from the DSM-IV.

The DSM-IV describes criteria for specific disorders. For example, the following is an excerpt from the DSM-IV that establishes the criteria for a diagnosis of autistic disorder (American Psychiatric Association, 1994, 70–71).

299.00 Autistic Disorder
A. A total of six (or more) items from (1), (2), and (3), with at least two from (1), and one each from (2) and (3):
   (1) qualitative impairment in social interaction, as manifested by at least two of the following:
      (a) marked impairment in the use of multiple nonverbal behaviors, such as eye-to-eye gaze, facial expression, body postures, and gestures to regulate social interaction
      (b) failure to develop peer relationships appropriate to developmental level

(c) a lack of spontaneous seeking to share enjoyment, interests, or achievements with other people (e.g., by a lack of showing, bringing, or pointing out objects of interest)

(d) lack of social or emotional reciprocity

(2) qualitative impairments in communication, as manifested by at least one of the following:

(a) delay in, or total lack of, the development of spoken language (not accompanied by an attempt to compensate through alternative modes of communication such as gesture or mime)

(b) in individuals with adequate speech, marked impairment in the ability to initiate or sustain a conversation with others

(c) stereotyped and repetitive use of language or idiosyncratic language

(d) lack of varied, spontaneous make-believe play or social imitative play appropriate to developmental level

(3) restricted, repetitive, and stereotyped patterns of behavior, interests, and activities as manifested by at least one of the following:

(a) encompassing preoccupation with one or more stereotyped and restricted patterns of interest that is abnormal either in intensity or focus

(b) apparently inflexible adherence to specific, nonfunctional routines or rituals

(c) stereotyped and repetitive motor mannerisms (e.g., hand or finger flapping or twisting or complex whole-body movements)

(d) persistent preoccupation with parts of objects

B. Delays or abnormal function in at least one of the following areas, with onset prior to age 3 years: (1) social interaction, (2) language as used in social communication, (3) symbolic or imaginative play.

C. The disturbance is not better accounted for by Rhett's disorder or childhood disintegrative disorder.

(Reprinted with permission from *Diagnostic and Statistical Manual of Mental Disorders*, Fourth Edition. Copyright 1994 American Psychiatric Association.)

Note that the diagnosis depends on various factors. Clients do not have to meet all the criteria to receive the diagnosis; this system allows for individual manifestations of the diagnosis. Since the DSM-IV is a way of classifying all types of mental disorders, most agencies have a copy of the DSM-IV.

The Diagnosis section of the report may be followed by a Prognosis section— a statement about future behavior. The Recommendations conclude the report and suggest some possible courses of action that would be beneficial in the psychologist's opinion, based on the psychological evaluation. For an example of a psychological report, see Figure 7.3.

Psychological evaluations differ according to the client's needs. The client profiled in Figure 7.3 was referred for assessment of his reading problems and to determine his eligibility for special services. The tests administered and the final report would be different if the client had been referred for other reasons (e.g., behavioral problems).

CONFIDENTIAL PSYCHOLOGICAL REPORT

NAME: Scott Garrett
AGE: 7 years, 6 months
DATE OF BIRTH: 3/2/XX
GRADE: 2
SCHOOL: Pineview Elementary
PARENTS: Mr. and Mrs. Scott N. Garrett
EXAMINER: Claudia Zimmerman
DATES OF ASSESSMENT: 9/30/XX

Reason for referral and background information
Scott Garrett was referred by his mother, Ms. Sue Garrett, who notes that
Scott doesn't enjoy reading and isn't good at it. She notes that he doesn't
appear to invest in classwork, particularly seatwork. His teacher, Ms. Cole,
is also concerned about Scott's progress. She notes that he is behind in
reading but is on grade level in math. According to the developmental
history, Scott accomplished developmental milestones in a typical fashion,
with the exception of speech. His first words were at about 18 months.
There was no history of physiological problems except for an eye muscle
imbalance problem; he is wearing corrective lens for that problem. His
hobbies are typical for his age. Birth history is unremarkable. Labor and
delivery were normal.

Assessment procedures
Wechsler Intelligence Scale for Children III (WISC-III)
Woodcock–Johnson, Revised, Tests of Achievement
Brigance Comprehensive Inventory of Basic Skills

Test behavior
Scott was evaluated during two sessions. In both sessions, he was willing
to work but would often ask, "Are we finished yet?" His problem-solving
attack skills seemed typical for a child his age, and he displayed no signs
of hyperactive or excessive distractibility in the one-to-one sessions. When
confronted with tasks requiring reading, he would sometimes exclaim,
"Oh, no!" In appearance, Scott is approximately average in height for his
chronological age, but somewhat overweight. He is an engaging

(continues)

Figure 7.3   Psychological report

youngster, with close-cropped blond hair and round facial features; he could be described as "cute."

Test results
On the WISC-III, Scott obtained a Full Scale IQ of 106 + or –3, a Verbal Scale IQ of 100, and a Performance IQ of 112. His Full Scale score of 106 is at the 66th percentile rank and is slightly above average for his chronological age. Chances are 2 out of 3 that the range of scores from 103 to 109 contain his true score. (A true score is the hypothetical average score a child would obtain on repeated testing with the same instrument, minus the effects of practice, fatigue, etc.). Scott's individual subtest scores are as follows. (Scores can be compared to a population mean of 10, a standard deviation of 3.)

VERBAL-SCALED	SCORES	PERFORMANCE-SCALED	SCORES
Information	11	Picture completion	11
Similarities	8	Coding	15
Arithmetic	12	Picture assessment	10
Vocabulary	10	Block design	11
Comprehension	9	Object assembly	12
Digit span	(13)	Symbol search	14
Mazes	12		

Scott's profile can be presented using a number of "factor scores." These factor scores have the same psychometric properties as a Full Scale score. That is, the population mean is set to 100 and standard deviation to 15 for the factor scores. The factor scores include Verbal Comprehension, 98; Perceptual Organization, 107; Freedom from Distractibility, 115; Perceptual Speed, 124. These scores reflect average to above-average performance in general, but there is some variability. For example, the Verbal Comprehension scores are considerably lower than the Perceptual Speed score and the Freedom from Distractibility score. In general, his Verbal Fluency, fund of general vocabulary words, and Verbal Reasoning and Judgment appear to be about average compared to chronological age-mates. He is able to focus attention when directed and maintain that attention and concentration to a degree significantly better than chrono-

Figure 7.3   *(Continued)*

logical age-mates. His nonverbal reasoning and synthesis/analysis scores appear to be average to slightly above average compared to age-mates. This intellectual profile is typically predictive of average-to-better classroom performance.

On the Woodcock-Johnson Revised Tests of Academic Achievement, the following scores were obtained:

	GRADE EQUIVALENT	SCALE SCORE	PERCENTILE RANK
Letter and word identification	1.6	94	35
Passage comprehension	1.7	97	43
Word attack	1.6	96	39
Reading vocabulary	1.7	95	37
Math calculation	2.8	97	43
Applied problems	1.8	97	43
Science	3.1	111	76
Basic reading skills	—	97	42
Reading comprehension	—	98	44

In general, Scott's scores on the Woodcock-Johnson ranged from slightly below average to slightly above average. His math- and science-related scores were slightly better than the reading scores. His language arts/reading abilities are approximately one grade level below his current grade level, which is somewhat consistent with teacher observations. His math scores were approximately grade appropriate, although applied problem-solving skills are depressed relative to straightforward calculation. Applied problems are compounded by a language component, which is not his forte. His best performance is in science, which he acknowledges as his "best subject." This suggests that Scott's relatively poor performance in language arts and reading may be motivational. Because his birthday comes late (September), he started the first grade young compared to other first-graders. Consequently, early language arts acquisition may have been particularly difficult for Scott, and there may be negative affect associated with reading skills currently. It should be mentioned that Scott's scores on the Woodcock-Johnson were compared to his chronological age-mates, not grade-mates. His scores would have been reduced by approximately 4 to 7 standard score points had age norms been applied.

(continues)

Figure 7.3   *(Continued)*

Results from the Brigance reveal some specifics associated with Scott's language arts problems. That is, he missed 3 words out of 10 on the primer level, 2 out of 10 on the grade 1 level, and 5 out of 10 on the grade 2 level. Scott seemed particularly adept at calling beginning consonants but often added a sound to the consonants. For example, rather than "mmm" for M, he responded with "mu." The same was the case for the letters B, H, J, G, R, S, D, M, and F. His greatest difficulty occurred with ending sounds. For example, RIX was pronounced as RIC and LIN as LINE. Other examples of ending problems included the following: SAT for SIB, TIDE for TID, PEN for PIN, TOX for TAX, and OX for OC. Obviously, Scott needs considerable work to master basic word lists.

Summary and recommendations
Although Scott is in the second grade, his language arts and reading skills are not consistent with that grade placement. Performance on the WISC-III indicates average to above-average intellectual ability. Results from the Woodcock–Johnson are consistent with classroom performance, which reflects relatively poor language arts and reading skills, but relatively stronger math and science skills. His medical history is normal, and there are no obvious physiological impairments. (The eye muscle imbalance apparently is being corrected by glasses, and exercise as prescribed by the physician should be continued.) Results from various tests suggest good reasoning and judgment, ability to concentrate and sustain atten-tion, and average nonverbal reasoning and judgment. One purpose of the evaluation was to rule out the presence of a developmental reading disorder, which would be a Diagnostic and Statistical Manual IV, Axis I 315.00 diagnosis of Reading Disorder. There is not sufficient evidence to warrant this diagnosis. However, the evidence does seem to be compati-ble with a general developmental delay. Scott started the first grade at a disadvantage relative to other first-graders. That is, his birthday comes in September and, in general, boys develop at a slightly reduced rate rela-tive to girls. Consequently, it is likely that his reading difficulties began because of developmental immaturity relative to age-mates, which is possibly confounded by early muscle imbalance problems of the eyes. It is possible to rule out the presence of specific learning disabilities/dyslexia, poor educational environment, low intelligence, and visual-auditory pro-cessing problems. The following recommendations are offered.

Figure 7.3 *(Continued)*

1. Scott's to-be-learned material should be individualized as much as possible to produce maximum gain and maximum motivation. Later, low-vocabulary/high-interest reading material can be assigned and a contract system developed to maximize reading and development automatized reading skills.

2. Because Scott's word-calling skills are poor, he should practice basic sight words, those most common in the reading content for his grade placement.

3. Scott would profit from instruction from a tutor. To-be-learned material should be coordinated with his classroom teacher.

4. Retention is not an appropriate option for Scott. There are considerable negative implications, primarily social, interpersonal, and self-esteem related. In addition, the literature is controversial regarding long-term academic gains associated with retention. Tutoring is a more appropriate solution. Tutoring will likely be phased out during the fourth and fifth grades.

5. Scott would profit from having stories read to him. Any activity that would increase his desire to read independently is appropriate.

6. Scott is overweight, and a low-calorie diet would be beneficial. A children's weight-loss program is available at University Hospital.

Claudia Zimmerman, Ph.D.
Licensed School Psychologist

Figure 7.3   (*Continued*)

Social History

For a complete case file, the client's past history and present situation must be investigated. The person's past adjustment can give indications of how he or she will adjust in the future. A **social history** also provides information about the way an individual experiences problems, past problem-solving behaviors, developmental stages, and interpersonal relationships. Some of the information in a social history may duplicate what has been gathered during the intake interview. In the social history, however, the client can relate the story in his or her own words, with guidance from the helper.

A social history has a number of advantages. Often the informal history taking leaves gaps, but the carefully done social history completes the picture. The case manager can then plan the appropriate integration of services and provide better information for future referrals. The social history often includes a better assessment of the client's need for services; this is especially helpful for clients who have multiple problems. A social history can also fulfill legal requirements. Finally, the process of taking a social history can help build the relationship between the case manager and the client.

There are also limitations to the social history. History taking is a preliminary activity in case management, but the client may perceive it as a phase in which solutions are put in place. Unfortunately, categorizations and judgments made at this stage may be premature. The process of taking the history can also give an inaccurate view of what will happen between the client and the case manager. Excessive questioning by the case manager may lead to a dependent role for the client, and culture-bound questions can create barriers to the development of the helping relationship. In addition, an exhaustive history is not absolutely necessary to develop a plan of services; it may be helpful, but the information gathered may not be relevant to service delivery. Spending too much time on history taking can also be harmful. The client may use the process to resist significant facts. Other clients may construe it as therapy, but it is not intended as such and may not even be therapeutically valuable. Despite these limitations, the social history still has the important function of completing the case file. Moreover, the case manager can use certain strategies to mitigate the limitations.

The following suggestions are adapted from McGowan and Porter's guide to history taking in rehabilitation settings (1967, p. 74). As phrased here, they apply to the gathering of information for a social history in other settings as well.

1. Do not let completion of the survey form or social history become a goal. Remember that the client is the main concern, not the paperwork.

2. Do not divert the client from discussing some aspect of his or her history because you already have the information necessary for the form and are anxious to move on to the next topic. Important clues to understanding a client may be gained from spontaneous discussion of the aspects of the history that he or she considers significant.

3. Be sure that the client understands the reasons for the history and can see the benefits of compiling it.

4. If a client resists part or all of the data gathering, or feels threatened by what may be perceived as an invasion of privacy, do not become defensive and punitive.

5. When additional information is deemed necessary, secure the client's permission to contact relatives, employers, friends, and others.

6. Do not let the client ramble on about all aspects of the history; maintain control of the interview.

Using these guidelines, the case manager gathers pertinent information about what appears to be the client's problem. The primary source of information is the client, who is encouraged to tell the story in his or her own way. The helper listens carefully to what is said, how it is said, and what is not said. The sequence of events, reactions, feelings, and thoughts are all taken into consideration as the client relates the history. Note taking should be kept to a minimum so that important nonverbal information is not missed.

Social history is taken within the context of the culture of the client. For example, interviews with individuals who belong to a collectivist culture must be treated with cultural sensitivity. In a collectivist culture, the focus is on the importance of the group rather than the individual. In a collectivist context, individuals must fit into the group; there is a focus on group values, beliefs, and needs; and the group influences individual behavior.

Because of group influences, social history may hold very different meaning to an individual from a collectivist culture than it would to a person in the American mainstream. As the client responds to questions and tells his or her story, there may be much more emphasis on the family and the community. The client may not be able to clearly define personal characteristics or personal problems, but describe them in terms of the group or family. It may appear that the client is avoiding answering the questions or not taking responsibility for his or her own behavior, but the client's experience of history may be that of the group or the family. It is also possible that the client may not wish to share his or her story. In many collectivist cultures, this information stays in the family or in the group.

There is no set form or procedure for taking a social history. Some agencies use forms to guide information gathering, such as the social data report shown in Figure 7.4. Others just provide guidelines for their case managers so the length and detail of social histories may vary. In all cases, the social history is prepared when a comprehensive picture of a client's situation is desired. The outline for writing it depends on what the agency wishes to emphasize, but certain topics are almost always included: identifying data, family relationships, and economic situation. Which other areas are emphasized depends on the focus of the agency and the presenting problem. For example, a social history of a couple involved in marital counseling might target such areas as family relationships and psychosocial development. For someone seeking economic assistance, important areas

SOCIAL DATA REPORT

Student's full name: _____ SS#: _____

Address: _____ Phone: _____

Age: _____ D.O.B.: _____ Race: _____ Sex: _____ Hair: _____

Height: _____ Weight: _____ Eyes: _____

Distinguishing marks: _____

Offense: _____

Date of offense: _____ Prior court record: _____ Yes _____ No _____
          (If Yes, show details under Additional Information.)

School attending:_____ Grade: _____

Have you been suspended from school, given detention, in-school suspension, or had truancy problems? Yes _____ No _____ (If Yes, show details under Additional Information.)

Health: _____ Are you on any prescription medications?

Yes _____ No _____ (If Yes, give type, amount, reason under Additional Information.)

Are you seeing or have you seen a mental health counselor? Yes _____ No _____

If Yes, where and when? _____

Have you ever used drugs or alcohol? Yes _____ No _____

Type, amount: _____

Are you employed? Yes _____ No _____ Where & hours: _____

Use of prescription medication: Yes _____ No _____

If Yes, type and amount: _____

Figure 7.4   Social data report

Parents' marital status: Living together, divorced, separated, widow(er)

If separated or divorced: When? _____

Discipline in home: Type: _____

Does it work? _____

Who provides discipline: mother, father, both, neither

Children in family:

Name　　Age　Lives at Home? (Yes or No)　Court/Arrest Record (Yes or No)

_____

_____

_____

_____

Are you receiving any financial assistance from anyone?

Type:　Food stamps　　　Amount:

　　　Social Security　　Amount:

　　　Child support　　　Amount:

　　　Other　　　　　　Amount:

Do you have medical insurance? (Name, policy #, address, and phone)

ADDITIONAL INFORMATION: _____

_____

(continues)

Figure 7.4　*(Continued)*

PARENT INFORMATION

Father's name:_____   Age: _____

Address: _____   Phone: _____

Employment: _____   Phone: _____

Educational level: 1   2   3   4   5   6   7   8   9   10   11   12   13   14   15   16GED

Court record: Yes _____ No _____ If Yes, for what and when? _____

Use of drugs or alcohol/Type of Use: excessive, moderate, little, none

Have you had any type of counseling?   Yes _____ No _____

If Yes, when and where? _____

Use of prescription medication: Yes _____ No _____

If Yes, type and amount? _____

If deceased, give date and cause: _____

Mother's name:_____   Age: _____

Address: _____   Phone: _____

Employment: _____   Phone: _____

Educational level: 1   2   3   4   5   6   7   8   9   10   11   12   13   14   15   16GED

Court record: Yes _____ No _____ If Yes, for what and when? _____

Use of drugs or alcohol/Type of Use: excessive, moderate, little, none

Have you had any type of counseling?   Yes _____ No _____

If Yes, when and where? _____

If deceased, give date and cause: _____

Figure 7.4   (Continued)

might be financial status, income, expenses, and work history. In general, the following areas may appear in a social history.

Identifying information: Name, address, date and place of birth, Social Security number, military service, parents' name and address, children's names and ages.

Presenting problem: Brief description of the problem.

Referral: Source and reason.

Medical history: Relevant hospitalizations, illnesses, treatment, and effects. Written permission is needed to obtain copies of medical records, if necessary.

Personal/family history: Family life, discipline, parenting, and personal development.

Education: Highest grade completed, progress, records.

Work history: Training, type and length of employment, ambitions.

Present family relationships and economic situation: Family members, ages, relationships, lifestyle, and income.

Personality and habits: Interests, disposition, social activities, personal appearance.

The client provides most of the information for a social history, but other sources may also contribute. When the case manager has gathered material from sources other than the client, it should be inserted under the appropriate headings, with the source identified. Direct knowledge is the main source, as in the following examples.

- She did not come for her first appointment.
- The client drummed his fingers on the table throughout the interview.
- He states that his goal is to receive a high school diploma and get a job.
- The client stated that during the past week she and her husband had three fights.

The next examples are statements of information from other sources.

- Educational records indicate that the client completed the sixth grade in school.
- Her parents report that the client lived with them until her marriage two years ago.
- He was fired from his job for absenteeism.
- A psychological evaluation indicates a mildly retarded 13-year-old with a possible hearing loss.

The social history shown in Figure 7.5 combines two approaches. The Identifying Information section is a form that the case manager completes. The remaining sections are a narrative based on information compiled from several sources (listed at the end of the report). At this agency, a social history may be compiled by more than one case manager, and all who are involved in the writing of the social history sign the written report.

SOCIAL HISTORY

I. *IDENTIFYING INFORMATION*

Date: 0/0/XX   Name: Joe Billy Smith   Date of birth: 0/0/XX

SS#: 000-00-0000   TennCare/Insurance provider: Blue Cross/Blue Shield

Policy number: 000000000   Marital status: Never married

Number of children: None reported

Address: 1111 Dogwood Drive

   Atlanta, GA

Telephone number: (123) 777-7777

Prevention resource: Community Health Center

Resource contact: Jim Therapist

County court:

Custodial dept.: Department of Youth Development

Custodial dept. contact: Jerry Officer

Date of state custody: 000000   County court: Cobb

Age: 16 years   Race: Caucasian   Height: 5' 2"

Weight: 120 pounds   Eye color: Blue   Hair color: Blond

Unusual markings: Tattoo on left bicep

Allergies: Penicillin   Current medications: None reported

Current medical problems: Ingrown toenail

Figure 7.5   Social history

Special circumstances: Ms. Smith (mother) reported that Joe was abducted from school at gunpoint and was kidnapped for five days. During the time Joe was gone, he was allegedly emotionally and physically harassed and assaulted. This took place the first two weeks in March.

II. *PRESENT PROBLEMS* (Current charges with dates and circumstances)

Joe appeared in Juvenile Court the following February on the charges of violation of a valid court order and disobeying his probation by using cocaine and alcohol. Joe pled true to these charges and was placed in the custody of the Department of Youth Development. On this same date, the court ordered an assessment of Joe and his situation.

III. *PREVIOUS PROBLEMS* (Past charges, adjudications, and placements)

According to Joe's court file, Joe was petitioned to court in October for the charge of running away from home. Joe pled true to the charge and was placed on probation with the County Probation Service. Ms. Smith (mother) also reported that when Joe disappeared in April, she went to the Police Department to file a missing persons report. Ms. Smith stated that she was informed that a missing persons report could not be filed, but that she could sign a paper declaring Joe a runaway juvenile. This action would allow the police to search for Joe. Ms. Smith reported that she signed the form only out of concern for Joe, and that this is now in Joe's court record. No charges were ever brought against the 20-year-old male who allegedly abducted and abused Joe.

IV. *FAMILY HISTORY* (Name, social security number, current address, phone, date of birth, marital status, employment, educational level, court record, alcohol and drug problems, mental and physical health problems, possible placement resource)

A. Father: Tom Smith is Joe's biological father. His date of birth is 0/0/XX, and his social security number is 000-00-0000. Ms. Smith reported that Mr. Smith lives in Dalton, Georgia, but she does not know his address. Ms. Smith stated that Mr. Smith's phone number is (123) 000-0000. Ms. Smith reported that Mr. Smith is currently remarried and is employed delivering bottled water. Ms. Smith does not know Mr. Smith's delivery route. Ms.

(continues)

Figure 7.5 *(Continued)*

Smith reported that Mr. Smith obtained his GED and has no prior court history, to her knowledge. Ms. Smith stated that Mr. Smith used to drink alcohol frequently—five times per week—but he did not use any drugs. Ms. Smith is unaware of any physical or mental health problems that Mr. Smith may have. Mr. Smith would not be an appropriate placement for Joe because of his sporadic interest in Joe's life. Ms. Smith stated that Mr. Smith has let Joe down many times in the past.

B. Mother: Betty Smith is Joe's biological mother. Her date of birth is 0/0/XX, and her Social Security number is 000-00-0000. Ms. Smith's current address is 1111 Dogwood Drive, Atlanta, Georgia, and her phone number is (123) 777-7777. Ms. Smith is divorced and works at Kroger's in South Atlanta. Ms. Smith reported that she completed the 10th grade and then earned her GED. Per Ms. Smith, she has no alcohol or drug problems or court history. Ms. Smith stated that she deals with mild depression and seeks professional mental health services when depression sets in. At this time, Ms. Smith reports that she is taking no prescription psychotropic medication for her depression.

C. Stepparents: Not applicable.

D. Siblings:

SIBLING NAME	GENDER	DOB	FULL/HALF SIBLING
Jeff Smith	Male	0/0/XX	Unknown
Ty Smith	Male	0/0/XX	Full

Sibling interaction:

Ms. Smith stated that Joe gets along well with his older siblings, but he is closest to Ty. Ty's wife has a baby girl, and Joe likes to help take care of her. Joe's siblings are older than he is, and Ms. Smith reports that Joe does not get to see them as much as she would like.

E. Other: (Grandparents, relatives, boyfriends, girlfriends, etc.)

Joe's last remaining grandparent died prior to this first introduction to the court. Ms. Smith reported that Joe was very close to this grandparent. Currently, Joe reports no girlfriend involvement.

Figure 7.5   (Continued)

V. *FAMILY INTERACTION* (Family dynamics/relationships, current issues, financial resources, needs, risks, etc.)

Ms. Smith reported that she and Joe are very close. Ms. Smith appeared to be very protective of Joe and defended him and his actions. Ms. Smith stated that the only current issue in the home is Joe's opposition to house rules. Ms. Smith is angry about Joe's placement in the Department of Youth Development, and she is threatening to sue the state. Ms. Smith reported that she earns minimum wage and usually works at Kroger's 30–40 hours a week. Ms. Smith stated that she pays rent and utilities. Ms. Smith reported that she receives food stamps, but she is not sure how much she will be getting now that Joe has gone into custody. Ms. Smith did not feel that all of her needs were currently being met.

VI. *HOME AND NEIGHBORHOOD* (Date of home visit, type of home, adequacy of space, housekeeping standards, hazardous conditions, neighborhood description)

A home visit was made by the case manager on 0/0/XX. Ms. Smith said they have been renting their two-bedroom apartment since July, prior to Joe's trouble, and that both she and Joe have plenty of room. This case manager observed that housekeeping standards were very good. No major inappropriate housing conditions were noticed. Their apartment is in a low-level crime area of Atlanta.

VII. *CHILD*

A. Early development history (Problems with pregnancy or delivery, planned/unplanned pregnancy, parental A&D use during pregnancy, developmental milestones, serious illnesses or accidents, diagnosis of hyperactivity, etc.)

Ms. Smith reported that she had no problems with the pregnancy or delivery of Joe; she delivered Joe herself, with the help of a midwife. Ms. Smith reported that she did not use alcohol or drugs while she was pregnant, and that Joe reached all of his appropriate developmental milestones earlier than most children. Ms. Smith reported that the only serious illness Joe had as a child was pneumonia.

(continues)

Figure 7.5   (Continued)

B. Peer interaction (Relationships with peers, age of friends, activities of friends, does or does not have friends)

Ms. Smith reported that Joe does not have many age-appropriate peers. Joe's friends are usually older. Per Ms. Smith, Joe does not do much in his free time except sleep. Ms. Smith was not certain what Joe does when he goes out with his friends, but she stated that he does not go out often.

C. Education (Last school attended, grade level, major school problems; accelerated, remedial, or special ed classes, truancy history)

Joe was last enrolled at Greenbriar High School in the ninth grade. Ms. Smith stated that Joe attends regular classes. Joe has had some truancy and tardiness problems. Since the abduction from school in April, Joe has become increasingly paranoid about attending school. It takes full cooperation with the school system to make him attend.

D. Psychological (Current and prior psychologicals, stating examiner's name, location of testing, and test dates)

Joe had a psychological evaluation, conducted at Lakeside Mental Health Institute in April. The examiner was Dr. John Doe.

Figure 7.5 *(Continued)*

VIII. *AGENCY CONTACTS AND SOURCES OF INFORMATION*

Name: Tom Casemanager   Relationship: Case Manager

Address: 201 Center Park Drive, 1100

          Atlanta, GA 12345

Phone: (123) 000-0000

Agency name: Lakeside Mental Health Institute

Agency contact: Dr. John Doe

Address: 5900 Lakeside Drive

          Atlanta, GA 12345

Phone: (123) 000-0000

Agency name: Department of Youth Development

Agency contact: Jerry Officer

Address: 222 Jail House Drive

          Atlanta, Georgia 12345

Phone: (123) 000-0000

Prepared by:

(Name)	(Title)	(Date)
(Name)	(Title)	(Date)
(Name)	(Title)	(Date)
(Name)	(Title)	(Date)

Figure 7.5   *(Continued)*

REPORT FOR JUVENILE COURT

Child: Lydia Maza, date of birth 0/0/XX   Petitioner: Jorja Mitten

Address: 100 Washington Pike,   Address: 100 Washington Pike,
        Chicago, IL   Chicago, IL

Mother: Leyla Mitten   Father: Lloyd Maza

Address: Unknown   Address: P. O. Box 18
                                           Hot Springs, AR

REFERRAL

The petitioner has been the primary caretaker of the child for ten years. The legal custodian is incarcerated in Shelby County Jail at this time. Petitioner asks that legal care and custody be given to her and her husband.

CIRCUMSTANCES OF THE CHILD

Lydia came to Chicago when she was 10 years old, to visit her maternal grandparents. During this time Leyla Mitten was arrested for drug trafficking, possession, and dealing and sent to Shelby County Jail, where she will be eligible for parole in two years. Jorja Mitten received a letter from her daughter asking her to care for Lydia until she is able to do so.

When asked if she would like to go back to Arkansas, Lydia stated that she would rather stay in Chicago. Lydia is a shy girl who states that she has more friends in Hot Springs, but has friends in Chicago too and enjoys playing with them. Lydia stated that she likes living with her grandparents, and she seems happy there.

Lydia's teacher at Ritta School says she is doing very well in class but seems emotionally fragile. The school records show that Lydia has missed four (4) days of school this year, all of which have been excused.

CIRCUMSTANCES OF PARENTS

Mother—Leyla, date of birth 0/0/XX, was arrested in Chicago when she was 25 on several drug charges. She will be eligible for parole in two

---

Figure 7.6   Court report

Another way social information appears in a case file is illustrated by the court report shown in Figure 7.6. It was prepared for juvenile court, based on social information gathered by a caseworker at the Department of Human Services (DHS). DHS caseworkers frequently prepare court reports, for example, if parental rights are being terminated or if the court asks DHS to investigate a petition for custody. All juvenile court reports have certain things in common, such years. Reportedly, Leyla expects Lydia to be returned to her then. Leyla has alcohol and drug issues.

Father—Lloyd Maza is a spa owner in Hot Springs. He has had no contact with Lydia since she was 10 years old.

CIRCUMSTANCES OF THE PETITIONER

Jorja Mitten, maternal grandmother, age 58, is the petitioner in this matter. She has been married to Gus Mitten, age 59, for 30 years. The Mittens live in a three-bedroom, one-bath home, which they own and have resided in for 18 years. The Mittens have two children, Leyla and her twin brother Boyd, who lives in Red Springs.

Ms. Mitten is currently employed by a local utility. She stated that she has worked there for 19 years. She reportedly earns $3,500 per month. Ms. Mitten denies any alcohol or drug abuse problems. She is a smoker, as is Mr. Mitten, and they use air filters in their home. Ms. Mitten states that she is in good health and takes Oruvail and Adalet daily under doctor's orders.

Mr. Mitten is currently employed. He stated that he has worked at Wind Industries for 22 years. He reportedly earns $4,000 per month.

Mr. Mitten denies any alcohol or drug abuse problems. He reports that he is in good health. Mr. Mitten takes Oruvail daily.

Mr. and Ms. Mitten are members at the YMCA, where they exercise weekly.

Ms. Mitten stated that she is concerned that Leyla is not emotionally or financially stable and is therefore in no position to care for Lydia. She stated that she is afraid that Leyla, after being released from jail, will "drag Lydia down" with her.

All references speak very highly of the Mittens and hold them in the highest regard. All stated that they have no concerns about the Mittens caring for Lydia.

A check with local law enforcement agencies revealed no prior record in Dade County on Jorja or Gus Mitten.

(continues)

Figure 7.6 *(Continued)*

as the reason for the referral to the department and the circumstances of the child, of both parents, and of the petitioner. Also included is the recommendation of the department, which the court may or may not follow. Although the format of this report is determined by the court, you will see content similarities to the social history in Figure 7.5. In this court report, a grandmother is asking for full custody of her granddaughter. A caseworker has been out to the home,

RECOMMENDATION

At this time, the department would recommend that custody be granted to Jorja Mitten. The legal custodian is currently unable to provide for the child, and the child states her reluctance to leave her grandparents' home. This worker knows of no reason why Lydia should not remain with Ms. Jorja Mitten.

As always, we will respect the court's wishes in this matter. We hope this information will be of assistance to the court.
Submitted by:

Illinois Department of Human Services

Tina Rachael
Caseworker 1

Figure 7.6   *(Continued)*

completed a social history of the family, and obtained a signed release of information from the petitioner. The caseworker has also consulted with the law enforcement agencies, checked references, and obtained as much information as possible from other sources. The caseworker then writes a report, informing the court as succinctly as possible of all the relevant information gathered.

Other Types of Information

Other types of information may be relevant to the case file, depending on the agency's mission and services as well as the client's problem. Educational and vocational information, the most commonly needed, is discussed here.

Educational information can have many parts: test scores, classroom behavior, relations with peers and authority figures, grades, suspensions, attendance records, and indications of academic progress such as repeated grades or advanced work. The sources of educational information are just as varied: school records, teachers, guidance counselors, and principals. Often, the particular information that the case manager obtains depends on which source is contacted. Rarely is it gathered in a single report, as medical information might be. In many cases, the case manager decides what information is needed and contacts the source or sources most likely to have that information. For example, a teacher is probably the best source of information about classroom behavior, whereas school records provide test scores and indications of past academic performance. The contact may occur formally (in writing) or orally (by telephone or personal interview).

Vocational information can be important for several reasons. People seem to be happiest when their activities are satisfying and fulfill their needs. There is also the need to earn a living, and self-support often engenders self-respect. Ways of gathering vocational information range from asking the client about his or her work history to arranging for a formal vocational evaluation. The types of information gathered include jobs previously held, the ability to get along with co-workers, work habits (e.g., punctuality and reliability), and reasons for frequent changes in employment. How much more information is needed depends on the client's problem and the agency's mission. For example, if the client has no work experience, an exploration of vocational interests and aptitudes may be in order. For the client who has had varied employment, the focus may shift to attitudes toward work and the skills developed. The client who has a substantial record may need help in reviewing his or her experience and skills to establish a vocational objective.

Let's return to Roy Roger Johnson's case, discussed in Chapter 1. Roy's counselor requested a period of vocational evaluation at a regional center that assesses an individual's vocational capabilities, interests, and aptitudes. Roy and the counselor, Tom Chapman, attended a staffing to hear the vocational evaluation report. Mr. Chapman later received a written report (Figure 7.7). The report illustrates two important points. First, information about a client is integrated with other new information to complete the picture. In this report, you will read about work history, medical information, and test scores, as well as the results of the vocational evaluation. Second, this report is a vocational evaluation report. Vocational evaluation is a process of gathering, interpreting, analyzing, and synthesizing all data about a client that has vocational significance and relating it to occupational requirements and opportunities.

Vocational and educational information add other dimensions to the client record, making the case file more complete. This information rounds out the case manager's understanding of who the client is—strengths, weaknesses, abilities, and aptitudes.

VOCATIONAL EVALUATION REPORT

ADMISSION DATA

To: Tom Chapman

From: Dan Howard

Re: Roy R. Johnson

D.O.B.: 0/0/XX

Date reported: 0/0/XX

Dates attended : 0/0/XX

Scheduled hours: 8:00 A.M.–3:00 P.M.
                              Monday–Friday

S.S.#: 000-11-2222

Sex: Male

Marital status: Single

No. of dependents: None

Date: 0/0/XX

Work history: Bartender, plumber's helper

Medication: None

Education: Two quarters SSCC

Transportation: Own vehicle

Program manager: Jo Singletary

TESTS ADMINISTERED

Purdue Pegboard

Valpar Component Work Samples

      Size discrimination

      Upper extremity range of motion

      Simulated assembly

      Eye–hand–foot coordination

      Bennett hand tool

Reason for referral: Mr. Johnson was referred to the Vocational Training Center for vocational evaluation to assess his manual and finger dexterity. This evaluation was to consist of paper/pencil testing and situational assessment. The client completed his paper/pencil testing, but did not complete his situational assessment. This report is based on the results of the tests he took and the limited time he spent in situational assessment.

BACKGROUND INFORMATION

The information in this section was based on statements made by the client on his evaluation interview.

    Social/educational: Mr. Johnson was a 30-year-old white single male who lived with his mother and sister. The client disclosed during his

Figure 7.7   Vocational evaluation report

evaluation process that his mother has been severely hearing impaired since birth. Because of this, he learned sign language before he could talk. He stated that he maintained this skill.

He had recently received a settlement of $40,000 due to his injury on the job and was building a home with these funds. He did not have any other source of income. The client completed two quarters at Silver State Community College. He quit when his grandmother became ill. He also received plumber's training for 6 months at T.A.T. and attended the Georgia School of Bartending for 5 weeks. The client has held 3 jobs in the past 3 years and has not worked since April. The client's stated vocational interest was in working as a mechanical engineer. He said he would choose this because it was within his physical capabilities, and that he had a background in mechanics.

WORK HISTORY

The client's work history was obtained from the client during his evaluation interview.

The client worked as a bartender at the Ramada Riverview for 10 months. He left to go to work at Joe B's Restaurant.

The client worked at Joe B's as a bartender for 2 months. He left this position because of a personality conflict.

The client last worked at Rock City Mechanical as a plumber's helper. He left following his injury.

MEDICAL INFORMATION

General: The client stated that he was in general good health. The client's general medical examination revealed that his only problems were related to his back injury.

Handicapping condition and functional limitations: The client was handicapped by a back injury while working on a plumbing job. It is stated in a report from Dr. Alderman that the client also had a history of previous lumbar disc disease. It is also stated that the client's recovery will be slow and that he will have a permanent impairment of 10% to the body as a whole.

(continues)

Figure 7.7   *(Continued)*

The client's limitations were taken from his general medical examination, done by Doctor Jones. These limitations include walking, standing, stooping, kneeling, lifting, reaching, pushing, and pulling.

Vocational implications: The client would work best in a sedentary job, one where he can get up and move around when needed to relieve his back pain. The client was still recovering, and these limitations may change with time.

BEHAVIORAL OBSERVATIONS

General: The client was cooperative during testing and attempted all tasks asked of him. In situational assessment, the client could not work on all the contracts that we had. He wanted to wait until we had something he was able to do. When we did not get work in that was easier than what we had, he quit coming in spite of efforts to make the work easier for him. The reason stated by this client for not attempting this work was that he feared injuring his back again.

Vocational: The client did not complete the situational assessment, and the observations made were not complete.

OBSERVATIONS OF WORK BEHAVIORS

The client was observed in 13 different areas while in situational assessment. Ratings were as follows:

Rating scale for work behavior	NA	Not acceptable
	1	Not acceptable; needs improvement
	2	Satisfactory; meets criteria
	3	Excellent; employment strength

Figure 7.7   (*Continued*)

Attendance	1
Punctuality	2
Co-worker relations	3
Supervisor relations	2
Work quantity	
Contract	3/1
Cleaning	NA
Work quality	3
Work tolerance	1
Job flexibility	1
Follows work instructions	3
Use of work time	2
Works without close supervision	3
Observes all safety procedures	3
Care of tools/equipment/materials	3

Comments

Attendance: The client attended the center 43% of the time. He did not come in because he could not do the work we had (sanding and painting dumpsters).

Work quantity/contract: The client's production on the nuts and bolts contract for Smithfield Industries was 95% of industrial norms. His production on the labeling contract for A.P.S.U. was 25% of industrial norms. It is noted that this contract required 100% quality.

Work quantity/cleaning: The client could not do this.

Work tolerance: The client could work a full day but complained of pain in his feet. He said that this was because of the concrete floor.

Job flexibility: The client could not do the work on the dumpsters or the cleaning duties. He worked on the A.P.S.U. labels, the tape-recycling contract for Triad Corporation of Tennessee, and the nuts and bolts contract for Smithfield Industries. These were all sedentary tasks.

(continues)

Figure 7.7   *(Continued)*

TEST INTERPRETATION

Purdue Pegboard
  Right hand—58%ile—industrial applicants
  Left hand—56%ile—industrial applicants
  Both hands—72%ile—industrial applicants
  Right+left+both—59%ile—industrial applicants
  Assembly—93%ile—industrial applicants

Valpar Component Work Sample #2—Size discrimination

	TIME	ERRORS
Assembly	115%	<100%
Disassembly	80%	100%

Performance—64%
Norm group: Seminole Community College Vocation Assessment Center

Valpar Component Work Sample #4—Upper extremity range of motion

	%ile	MTM%
Assembly: Dominant hand	85	70
Assembly: Other hand	80	75
Disassembly	70	90

Performance: 87%ile
Norm group: Seminole Community College Vocational Assessment Center

Valpar Component Work Sample #8—Simulate assembly
  Performance—95%ile   MTM—75%
    Norm group: Seminole Community College Vocational Assessment Center

Valpar Component Work Sample #11—Eye-hand-foot coordination
  Points Performance—25%ile   MTM—50%
  Time Performance—85%ile   MTM—140%

Bennett Hand-Tool Dexterity Test
The client's performance on this instrument was 75%ile when compared to boys at a vocational high school.

Figure 7.7 *(Continued)*

SUMMARY/RECOMMENDATIONS

Summary: Mr. Johnson was a 30-year-old man who was handicapped by a back injury, with limitations in walking, standing, stooping, kneeling, lifting, reaching, pushing, and pulling. The counselor is referred to the medical information section of this report for details.

The client's performance in testing could be considered strong, with the exception of his accuracy in performing the eye-hand-foot coordination work sample. The client's scores (except eye-hand-foot, as mentioned above) were above average when compared with the norm group.

The client's strengths in situation assessment were in co-worker relations, work quality—nuts and bolts contract, working without close supervision, observing all safety procedures, and care of tools/equipment/materials.

The client's weaknesses in situational assessment were attendance, work quantity—labeling contract, work tolerance, job flexibility.

The client had acceptable performances in punctuality, supervisor relations, and use of work time.

Recommendations: The results of the testing done here at the center to assess the client's dexterity show that he could perform a sedentary, small, hand-assembly job. He expressed no interest in pursuing this line of employment. As noted in the Medical Information section, he would need to be in a job that would allow him the freedom to get up and move around when he needed to relieve his back pain. During situational assessment, he was observed to have the need to get up and down as he worked.

As stated in the Social/educational section of this report, the client is proficient in sign language for the hearing impaired. The client could be employed as an interpreter for the hearing impaired if he wanted to pursue certification. The client said that he had never considered this, but was open to the idea. It is recommended that the counselor investigate the possibility of having the client obtain his certification.

(continues)

Figure 7.7 *(Continued)*

    This client expressed an interest in attending the state university to
pursue a degree in mechanical engineering. This evaluator cannot make
a recommendation regarding this pursuit, since no academic testing is
performed at this facility.

_____

Dan Howard, Evaluator

_____

Jo Singletary, Center Manager

Figure 7.7   *(Continued)*

Chapter Summary

The information about the client that is gathered from other professionals helps the case manager see a more complete picture of the client. This information includes medical reports, psychological evaluations, social histories, and educational and vocational information. When the case manager requests the information from other professionals, the goals must be clear, and it is helpful if the client's problems are identified. Once the information is received, the case manager reviews it and integrates the results with the information previously gathered.

Medical information is critical, especially when a client has disabilities or mental illness. It is important for the case manager to understand medical terminology and be familiar with medications. A psychological evaluation is also an important part of a client file because it contributes to the understanding of the client as an individual. Often the case manager needs a psychological evaluation to establish eligibility for services, to justify a service, or to screen for criteria to determine need for services

Social histories provide information about the way an individual experiences problems, past problem-solving behaviors, developmental stages, and interpersonal relationships. The social history can help to complete the picture about the client and assists the building of the relationship between the case manager and the client.

Other information, such as educational and vocational information, may be included in the file. Relevance is determined by the agency's mission and services, as well as the client's problem.

Chapter Review

◆ *Key Terms* ....

Medical diagnosis   Word root
Medical consultation   Psychological evaluation
Physical examination   Psychological report
Diagnosis   DSM-IV
Medical terminology   Social history

◆ *Reviewing the Chapter* ....

1. Identify the resources that will help you understand medical reports.
2. How does medical information contribute to a case file?
3. In what situations would a medical consultation help you?
4. Describe a general medical examination.
5. Define each of the elements that form medical words.
6. What are the three basic steps in working out the meaning of a medical term?
7. Why is keeping current with medical terms a challenge for case managers?

8. List reasons to refer a client for a psychological evaluation.
9. How does a case manager make a good psychological referral?
10. Describe a psychological report.
11. What is a social history?
12. Describe the advantages and limitations of a social history.
13. How will the guidelines for history taking help you do a social history?
14. Complete a social data report (Figure 7.4) on yourself.
15. Write a social history on yourself, using the nine content areas of a social history.
16. Describe the three ways in which a social history may appear in a case file.
17. What do educational and vocational information add to a case file?

◆ *Questions for Discussion* ....

1. Why do you think it's important to have medical information?
2. What difficulties do you expect to have in understanding a psychological report?
3. Develop a plan to gather information for a social history of a client who is in prison for armed robbery.
4. Do you believe that you can have too much information about a client? Why or why not?

References

American Psychiatric Association. (1994). *Diagnostic and statistical manual of mental disorders* (4th ed.). Washington, DC: Author.

Dabatos, G., Rondinelli, R. D., & Cook, M. (2000). Functional capacity for impairment rating and disability evaluation. In R. D. Rondinelli & R. T. Katz (Eds.), *Impairment rating and disability evaluated* (pp. 73–94). Philadelphia: W. B. Saunders.

Felton, J. S. (1992). Medical terminology. In M. G. Brodwin, F. Tellez, & S. K. Brodwin (Eds.), *Medical, psychosocial, and vocational aspects of disability* (pp. 21–33). Athens, GA: Elliott & Fitzpatrick.

McGowan, J. F., & Porter, T. L. (1967). *An introduction to the vocational rehabilitation process.* Washington, DC: U.S. Department of Health, Education, and Welfare.

Tallent, N. (1993). *Psychological report writing* (4th ed.). Englewood Cliffs, NJ: Prentice Hall.

Chapter Eight ....

Service Coordination

K *nowledge of community resources is a must, but that is something that you learn. And the little secrets on how to navigate, some of that you learn as you do it.*
>—Jan Cabrera, Intensive Case Management Services, Los Angeles, personal communication, March 24, 1998

A *s far as doing case management in the hospital, I saw in many ways that it was crucially important to have networks in support services and know the agencies because you certainly couldn't address all the needs of a client. I had to know who in detox I was sending my people to, how to go about getting Social Security, and to help families know how to deal with the process.*
>—Judith Slater, Kennesaw State University, Kennesaw, GA, personal communication, October 14, 1995

C *ommunication with other service providers is really important, knowledge of community resources. The case manager has to be a broker. He or she has to be able to work well with lots of other people.*
>—Alan Tiano, Pima Health System, Tucson, AZ, personal communication, October 7, 1994

One of the most important roles in case management is service coordination. Rarely can a human service agency or a single professional provide all the services a client needs. Because in-house services are limited by the agency's mission, resources, and eligibility criteria—as well by as its employees' roles, functions, and expertise—arrangements must be made to match client needs with outside resources. Case managers must have knowledge of available community resources and the skills to put them to use.

The preceding quotations reflect the knowledge and skills that a case manager uses to meet client needs. According to Jan Cabrera, the case manager at Intensive Case Management Services, knowledge of resources is important, as is negotiating the service delivery system to gain access to those resources for the client. A key to using resources, according to Judith Slater, is to have networks in place so that the case manager knows both the resource agency and the name of a contact. Perhaps the one indispensable skill in using resources is communication; Alan Tiano states that working well with other service providers depends on the case manager's communication skills.

Today's service delivery environment requires new roles and responsibilities for the case manager. In the past, many services were provided directly by the case manager, but service delivery has become more specialized. Profession-als must be careful not to provide direct services in areas in which they are not trained or lack the necessary resources. Case management has thus come to mean providing selected services, coordinating the delivery of other services, and mon-itoring the delivery of all services. This shift in job definition calls for skills in ad-vocacy, collaboration, and teamwork.

This chapter explores service coordination as a critical component of mod-ern case management. We examine the coordination and monitoring of services as well as the skills that will help you perform these roles. After reading the chap-ter, you should be able to accomplish the following objectives.

COORDINATING SERVICES
- Describe a systematic selection process for resources.
- Make an appropriate referral.
- Identify the activities involved in monitoring.
- List ways to achieve more effective communication with other professionals.

ADVOCACY
- Explain why advocacy is important.
- Name several client problems that are appropriate for advocacy.
- Identify ways to be a good advocate.

TEAMWORK
- Describe the purpose of a treatment team.
- Define departmental teams, interdisciplinary teams, and teams with family and friends.
- List the benefits of working in and with teams.

Coordinating Services

If a client needs services that an agency does not provide, it is the case manager's responsibility to locate such resources in the community, arrange for the client to make use of them, and support the client in using them. These are the three basic activities in coordinating human service delivery. In coordinating services, the case manager engages in linking, monitoring, and advocating while building on the assessment and planning that have taken place in earlier phases of case man-agement. The case manager continues to build on client strengths or emphasize client empowerment, continually aware of the client's cultural background and the client's basic values.

Coordinating the services of multiple professionals has several advantages for both the case manager and the client. First, the client gets access to an array of services; no single agency can meet all the needs of all clients. The case manager can concentrate on providing only those services for which he or she is trained while linking the client to the services of other professionals who have different areas of expertise and have the necessary resources. Second, the case manager's knowledge and skills help the client gain access to needed services. Often, services are available in the community, but clients are unlikely to know what they are or how to get them. The success of service delivery may depend on advocacy by the case manager. Also, service coordination promotes effective and efficient service delivery. In times of shrinking resources, demands for cutbacks in social services, and stringent accountability, service provision must be cost effective and time limited. In addition, customer satisfaction is important. Clients have a right to receive the services they need without getting the runaround or encountering frustrating confusion among providers.

Service coordination becomes key once the client and the case manager have agreed on a plan of services and determined what services will be provided by someone other than the case manager. For services that will be provided by others, a beginning step is to review previous contacts with service providers. What services do they provide? Is this client eligible for those services? Can the services be provided in-house? What about the individual's own resources and those of the family? Family support may be critical for the success of the plan, and the client's own problem-solving skills may be helpful. A thorough case manager does not ignore the resources of the client, the family, or significant others. The next step is referral—the connection of a client with a service provider. The final step is monitoring service delivery over time and following up to make sure the service has been delivered appropriately. These steps may vary somewhat, depending on whether the services are delivered in-house or by an outside agency, but the flow of the process is likely to be the same. Before examining these steps in detail, let's review the documentation and client participation aspects of service coordination.

Documentation is critical in this part of case management. Staff notes must accurately record meetings, services, contacts, barriers, and other important information. During this phase, reports from other professionals are added to the case file. Any progress that occurs in the arrangement of services must be recorded by the case manager.

Client participation is important throughout the service coordination process. This entails more than just keeping the client informed; his or her involvement should be active and ongoing. First of all, the client participates in determining the problem that calls for assistance. Second, the values, preferences, strengths, and interests of the client play a key role in selecting community resources, and of course client participation is critical in following up on a referral. Clients also have the right to privacy and confidentiality. Without the client's written consent, the case manager must not involve others in the case or give an outsider any information about it.

Resource Selection

Once client needs and corresponding services have been identified, the client and case manager turn their attention to **resource selection**—selecting individuals, programs, or agencies that can meet the needs. Paramount in this decision is consideration of the client's values and preferences. The information and referral system the case manager has developed (see Chapter 6) is useful in this regard.

Rube Manning is a 53-year-old white male who is on parole for aggravated rape. He had sexual relations with his 12-year-old niece; she later gave birth to his son. Both parties claim that the intercourse was consensual; the severity of the charge and conviction was due to the girl's age. The girl and the family seem to harbor no animosity toward Rube, even going so far as to write a letter on his behalf to the Department of Corrections. Rube was sentenced to three years in prison and is now eligible for parole. Angela Clemmons is the parole officer assigned this case. She and Rube must develop a plan of services for him to pursue once he is released. Among the conditions of Rube's parole are completing a mandatory sex offender program, supporting his son, and finding employment.

There are no options for the mandatory sex offender program; only one program is available in this community. Angela senses that Rube is motivated to do everything in his power to comply with the conditions of parole. Although he does not talk much about his prison experience, he does say that he didn't like it. Angela suspects that he was abused by other inmates. Sex offenders are usually on the lower rungs of the prisoner hierarchy unless they are very strong or charismatic; Rube is neither.

Finding employment and supporting the child are tied together. Checking her information and referral file, Angela advises Rube that there are three short-term training programs that can provide him with job skills. The first two are at the vocational school and would give him a certificate in either horticulture or industrial maintenance. The third one is on-the-job training in food services, with a modest salary until training is finished. Rube's preference is horticulture because he grew up on a farm and thinks he would feel more comfortable outdoors. He knows that industrial maintenance is a fancy term for janitorial work, and he's not interested. The location of the food services training is not on the bus line, and Rube has no transportation of his own, but this option offers a salary immediately. Angela notices that Rube sounds interested—even a little excited—about horticulture, so she checks her addresses and e-mail file for the phone number of her contact. (See Figure 8.1.)

In this case, resource selection has been systematic, which has advantages for both the client and the case manager. The client and the case manager proceed objectively and deliberately, taking into account Rube's values, beliefs, and desires. The rationale for the choice is articulated, and it reinforces his motivation to follow through with the referral. Rube Manning and his parole officer have chosen the horticulture program: It is on the bus line, it builds on Rube's previous farming experience, and it is something he wants to pursue.

The selection process can also accommodate many alternatives and can tailor services to the client's unique circumstances. The conditions of Rube's parole include work, and he does want the independence, salary, and respect that come with employment. However, he is not willing to do just anything. Being a janitor doesn't appeal to him, and he does not want to work indoors. Had the parole officer ignored his feelings at this point and decided to steer him toward janitorial

---

Agency: Lincoln Vocational Technical School

Address: 30512 Townview Parkway

Contact Person: Lynda Johnston, Admissions
Robert Griffin, Student Services

Phone: 555-1516

Services: Short-term training programs in auto mechanics, cosmetology, horticulture, industrial maintenance, printing, and secretarial services.

Comments: Good student services and advising: Janet Evans 7/1/XX

Figure 8.1   Entry in information and referral file

work, Rube would probably not be motivated to do well. At the very worst, he would do nothing, and his parole would be revoked. In addition, the relationship between Angela Clemmons and Rube Manning would not develop as a partnership. Instead, their decision to try the horticultural program takes into account Rube's wishes, along with his need for training and employment.

Being aware of the client's preferences, strengths, and values is critical to the success of the selection step in service coordination. There must be a strong partnership between the participants.

Making the Referral

As mentioned earlier, no helper can provide all conceivable services. Therefore, arrangements must often be made to match client needs with resources. This is done by referring the client to another helping professional or agency to obtain the needed services. Referral is the process that puts the client in touch with needed resources. According to Alex Monteiro, working with parents of students at Roosevelt High School, "we realize how many individuals there are in a family and a limited source of income and we refer them to services . . . we have some hotlines that we can refer them to and often the Department of Social Services. Of course I refer them to the churches, too. They have places for homeless and shelters and things like that" (personal communication, March 24, 1998).

A **referral** connects the client with a resource within the agency structure or at another agency. In no way does referral imply failure on the case manager's part. Limitations on the services a case manager can personally provide are imposed by policy, rules, regulations, and structure, as well as his or her own expertise or personal values.

The case manager assumes the role of broker at this point in service coordination. The broker knows both the resources available in the community and the policies and procedures of agencies. He or she acts as a go-between for those who seek services and those who provide them. Consider the following case with regard to the referral process and the broker role.

Bethany's first client on Tuesday is Anna, a young woman who has just discovered that she is pregnant. This pregnancy has caused a crisis in Anna's family. Her parents are first-generation immigrants from San Salvador, they are Catholic, and they are very opposed to both pregnancy and abortion. Although the agency that employs Bethany specializes in career development services, Anna feels comfortable with Bethany and wishes to discuss her options about the pregnancy with her. On the other hand, this is a difficult subject for Bethany because her sister had an abortion three years ago and still feels guilty and upset about her decision. In fact, the whole family is still having difficulty with it since the sister is living at home. Bethany also knows that her training is in career development, and she has never worked with anyone dealing with an unwanted pregnancy.

The encounter illustrates a situation that is appropriate for a referral. Bethany has some personal feelings that may impair her objectivity; she recognizes that she has no professional experience with this problem; and her agency's purpose is career development. For these reasons, she decides it is best to make a referral to someone who can help Anna explore options related to the pregnancy. Bethany will continue to support Anna's career development efforts. In the referral process, Bethany's role is that of a broker.

Making a referral may seem like a fairly uncomplicated process, but it often results in failure. If a case manager believes that all that is necessary is being aware of client needs and making a phone call, the referral is likely to be unsuccessful. In fact, it is common for clients referred to other community resources to resist making the initial contact. Clients may also fail to follow through after the first interview and drop out before service provision is complete.

A referral can fail for three reasons. The first is insensitivity to client needs on the part of the case manager. Identifying the problem but failing to grasp the client's feelings about it contributes to an unsuccessful referral. The client may not be ready for referral at this point, feeling only that he or she is being shuffled among workers or agencies. Second, if the case manager lacks knowledge about resources, the client may be referred to the wrong resource. This makes him or her feel lost in the system, think that it is all a waste of time, and believe (sometimes correctly) that the case manager is incompetent. A third reason for failure is misjudging the client's capability to follow through with the referral. Suggesting to an involuntary client that she call to make an appointment for a physical examination may not work, perhaps because she is new to town, is unsure who to call, doesn't have a phone, or may not even want this exam.

How can the case manager make the referral process a successful one? Two keys to success are assessing the client's capabilities and forming a clear idea of the referring professional's own role (Brill, 1998). Assessing clients' capabilities means finding out how much they can do on their own. It is good to encourage independence and self-sufficiency in clients, but some of them will prove unable to identify what they need and take the steps to obtain it. The nature of the problem, the feelings the client has about it, and the energy required for action may all contribute to feelings of being alone, an inability to act, and a lack of motivation to follow through.

In addition to assessing the client's capabilities, the referring case manager must form a clear idea of what role he or she will play in the referral process. In this, the case manager should be guided by what the client needs and what relationship the case manager has with the other professional or agency. The case manager's degree of involvement in the referral can fall anywhere on a continuum, from discussing several resources with the client, who then takes responsibility for selecting a resource and following through, to giving concrete assistance with details, such as making the appointment on the client's behalf and having an agency volunteer accompany him or her to the appointment.

Bethany approached the referral process in the following way. She acknowledged Anna's concern about her situation and recognized her desire for some help. She also shared with Anna her reservations about being able to assist her, explaining that her training was in career development and she had limited knowledge about options for an unmarried pregnant woman. However, she did know of two agencies that offered just the services Anna was seeking. Anna wanted to know about these, so they discussed the services they provide and their geographic locations. Anna was concerned about the cost of services, and Bethany was unsure about the agencies' charges. She checked her computer file and found that both agencies charged fees on a sliding scale. Anna didn't know what that meant, so Bethany explained that such a scale determined the fee in accordance with the individual's income. Anna was unsure how to get an appointment—who to call, how to explain the problem, and so forth. She also wondered whether she would be able to continue working with Bethany on career development. Bethany discussed all these concerns with Anna. Together, they decided on one of the agencies, and Bethany agreed to make the initial contact. Her previous work with Anna led her to believe that once the initial anxiety of making contact was over, Anna was capable of showing up for the appointment and getting the services she needed.

BETHANY: Hello. This is Bethany Douglas at Career Development. I am working with a client who needs help identifying her options for dealing with an unplanned pregnancy. Will someone at your agency see her?

RECEPTIONIST: We do provide counseling. Let me connect you with one of our counselors.

COUNSELOR: Hello, this is Carol Fong. May I help you?

BETHANY: Yes, Bethany Douglas here. I am a career counselor at Career Development. My client has just found out she is pregnant and would like to talk with someone about her options.

She is 19 and single. Could we set up an appointment for her to come see you?

COUNSELOR: Yes, I would be glad to see her. Would Monday at 11:00 o'clock be okay?

BETHANY: *(Checks with Anna, who nods)* Yes, that would be fine. Her name is Anna Rodriguez. She will see you at 11:00 o'clock Monday. Thank you.

Bethany used several strategies to ensure that Anna's referral was a successful one.

1. *Discuss with the client the services that are provided by the resource.* The discussion should include why the referral is needed, how it will be helpful, how the client feels about it, and what information should be provided. If client information will be shared, then a release form is signed by the client or guardian at this time.

2. *Make the referral.* Making a referral may entail no more than providing the client with a telephone number and an address; helping him or her with the initial contact, as Bethany did; or taking the initiative to contact the resource. The interaction may involve scheduling an appointment, telling what the client knows about the resource, and finding out what information the resource needs. Of course, before any information is released, the client's permission must be obtained.

Suppose that the referral did not go as planned. When Bethany made the call, Carol Fong might have responded differently—perhaps she couldn't possibly see Anna until next month, or her agency didn't do that kind of counseling anymore. Bethany would have two options. She could return to her file to locate another agency that provides the services Anna needs. However, suppose that Anna lives in a small town or a rural area where there aren't any other agencies to call. Bethany's second option would be to become a **mobilizer**—one who works with other community members to get new resources for clients and communities. Bethany could try to mobilize Carol Fong and other professionals so that needed services could be made available to Anna.

3. *Share the referral information with the client.* He or she needs to know the appointment time, the location, and the name of the person to see on arrival. It is also appropriate to find out what support the client might need to follow through with the appointment.

4. *Follow up on the referral.* The service coordinator can do this by talking with the client and the helper who received the referral. Did the client show up? What happened? Was the client satisfied with the services? With the worker? Service coordinators with thorough information and referral systems make a habit of noting such information in their files. Information from the worker who saw the client may be conveyed in a phone call, a written report, or not at all.

Bethany followed up on Anna's referral by talking with her about it the next time they met. She discovered that Anna had had no trouble finding the agency, liked the worker immediately, and felt positive about exploring her options with her. Bethany received no official report from the other agency and did not request one.

The referral process is a flexible one that can be adapted for use with any client, but client participation is vital to good service coordination. Clients participate in the decision to refer and the choice of where to refer. Their capabilities determine the extent of their involvement in the steps of the referral process, including making an appointment, getting to the agency, and so forth.

The case manager's role in the referral process varies from little involvement to integral involvement, depending on the client's capabilities. The case manager's responsibilities include knowing what resources are available for the client, how to make a referral, and how to assess the client's capabilities accurately. His or her involvement does not end after the referral; the next step, monitoring services, is also the case manager's responsibility.

Monitoring Services

Once the referral is made, monitoring service delivery becomes the focus of case management. Monitoring services is more than following up on the contact; it may mean offering information, intervening in a crisis, or making another referral. The case manager continues to work in the roles of broker and mobilizer throughout this phase of service coordination. In **monitoring services**, the case manager reviews the services received by the client, any conditions that may have changed since the planning phase, and the extent of progress toward the goals and objectives stated in the plan. This review can occur as often as once a day or three times a week or as little as once a month or once a year, depending on the goals of the program, caseload of the helper, and resources available.

REVIEW OF SERVICES

Once a case manager has made a referral, delivering the needed service becomes the responsibility of the resource—the agency or professional that has accepted the referral. However, the case manager does not relinquish the case completely. He or she remains in contact with the client to ensure that the services are being delivered, that the client is satisfied with them, and that the agreed-on time frame is maintained. As you remember, all these are specified in the plan of services.

When checking with Rachel Vasquez after her visit to the health clinic, the case manager heard about the generous time a volunteer had spent with Rachel in making out a balanced nutrition plan for her diabetic son. Rachel was excited about knowing what to buy, how to prepare it, and why it made for a good meal. Most of all, she was impressed by how much time the volunteer spent with her.

If there are problems with service delivery, the case manager has ultimate responsibility to intervene. Problems may be caused by the agency, the client, or both. For example, the agency may prove unable to see the client for several weeks, or may neglect to do what the client has been promised. The client, on the other hand, may fail to show up for appointments or refuse to cooperate (e.g., be reluctant to give needed information). The case manager must be aware of the situation if he or she is to know that intervention is required. The intervention in such a case involves identifying exactly what the problem is and working with the client and the resource to resolve it.

Sam Miller received a call from the VA hospital where 22-year-old Raymond Fields (who is mentally retarded) had been placed as an orderly just two weeks before. Both the supervisor and Raymond had been pleased with the match. This morning, the supervisor reported that twice in the past three days, Raymond had been seen unzipping his pants and playing with his penis in the hallways. Sam hastened over to the hospital to talk with Raymond about the behavior. He told Raymond to keep his pants zipped. There was no more trouble afterward.

CHANGING CONDITIONS

Often there is a time lag between plan development and the provision of services. During this period, the case manager seeks agency approval, if necessary, and arranges for services either within the agency or at another. It is also likely that there will be changes in the client's situation during this time. Living arrangements, relationships, income, and emotions are some of the factors that may change. The presenting problem may show some alteration, or additional problems may surface. Any such changes may necessitate review and revision of the plan.

Alma Justus is raising two granddaughters and one grandson with the help of her own son, Zack. The mother of the children, Alma's daughter, lives in another state with her boyfriend and his two kids. Alma and the children are receiving assistance from a case manager at the local Office on Aging. Last week Alma was placed in the hospital, and after extensive testing, it was confirmed that she had had a series of slight strokes. The case manager will work with the family to determine the changing need for services.

The client's circumstances may also change during service delivery. Part of service monitoring is staying informed of changes that occur in the client's life. Some changes may occur as a result of service delivery, for example, a client might learn more appropriate ways to express anger than hitting his spouse. Other changes may have nothing to do with service delivery, yet they influence it. For example, a client might decide to marry while halfway through service delivery—an action that could well affect her economic eligibility for services.

Again, monitoring of services helps the case manager stay abreast and be ready to intervene if necessary.

EVALUATING PROGRESS

Monitoring services also entails continually checking progress toward the goals and objectives set forth in the plan of services. Continual evaluation may lead to modification of the plan so as to improve effectiveness or deal with new developments. In monitoring services, the case manager repeatedly asks the following questions.

- Has the identified problem changed?
- Was the referral made correctly?
- Were the desired outcomes achieved?
- Should the plan be altered?
- Should the case be closed?

Monitoring of services goes most smoothly if close contact is maintained with the client. Outcome measures focus on the client, so he or she is a key source of information about service delivery. Did the client use the resource? Was the goal of the referral attained? The case manager's responsibility continues until the client's problem is resolved. Follow-up and monitoring are performed to make sure that referrals result in the desired outcomes.

The following case illustrates how a case manager monitors services by reviewing the services received, considering any changes in conditions, and evaluating progress toward goals and objectives.

It was chilly on February 17, but the Naylors were happy. It was Presidents Day weekend, and they were going to have three days off. Everyone gathered in the den in front of the fireplace. Jennifer, the younger daughter, was wearing a tank top and shorts to be comfortable, since she had just come down with chickenpox. Johanna, the older daughter, went to look for the kitten her grandparents had given her for Christmas. It appeared to be just another quiet evening.

Johanna came in the back door about 6:30 P.M. and said, "Mom, there's a fire in the garage!" Mrs. Naylor looked out the door that led to the garage and saw flames at least 10 feet high. Calmly she said, "Everybody out," and headed for the front door. All three of them made it out safely. As the Naylors stood watching the fire consume their home, they wondered what they were going to do and where they were going to go. Would they be able to salvage anything at all?

The Naylor family lost everything. The Burn Shelter in their community immediately stepped in to provide the many services that fire victims need. Bettyjean Fleming, a case manager at the Burn Shelter, was assigned to work with the Naylor family. Once notified of the fire, Ms. Fleming went to the site of the fire to help the family with their immediate needs. Comfort, clothing, a meal, transportation to the hospital, and temporary lodging are among the services the shelter provides. Case managers work with the victims to cope with any losses they may have suffered (including family members, pets, and possessions). Once the immediate needs are met, the counselor and the victims develop a plan to meet long-term needs, which will be met by other agencies.

The Naylors had a number of needs: transportation, housing, clothing, household furnishings, and counseling. One of the most serious needs was counseling for the two girls. Not only had they lost everything they owned, but they also lost every picture and memento of their father (who had died a year before). They desperately needed help in coping with the loss of their father, as well as loss of their home, their possessions, and their pet. Ms. Fleming learned of these circumstances from Mrs. Naylor, and they developed a plan for the long-term services. She was supportive of counseling for the girls; together, they discussed several alternatives, using the file of community services at the Burn Shelter. They decided on counseling services at the local mental health center, which is known particularly for its children's services. Mrs. Naylor agreed to call for an appointment. Unfortunately, she was told that the earliest appointment was next month; she didn't see any alternative, so she took that appointment. The following week, Ms. Fleming called to see whether Mrs. Naylor had received her insurance check. In the course of the conversation, she learned about the delay in getting counseling. She alerted Mrs. Naylor to the emergency services that were available and volunteered to call and talk with a counselor. Ms. Fleming often worked with staff at the mental health center, so she called one of her contacts, Frances Lane. Ms. Lane agreed to see the girls that very week and requested some background information, which was provided. Mrs. Naylor was grateful for the intervention since there had been no improvement in the girls' mental state.

At the next meeting with Mrs. Naylor, Ms. Fleming asked about the visit to the mental health center. Were they seen on time? How did the girls feel about Ms. Lane? Were they feeling better? What were the next steps? At the same meeting, Ms. Fleming asked again if the insurance check had arrived, how the apartment was working out, and whether Mrs. Naylor or the girls had other needs.

In this case, the case manager's responsibility did not end after the referral. Because she monitored service delivery, Ms. Fleming became aware that there were problems. She was able to intervene, using her contacts to make a difference in service delivery. She also monitored the delivery of the other services, verifying that the identified services were delivered within a reasonable time and by the professionals designated to deliver them. Not keeping abreast of developments in a case can result in delays in needed services (or even nonperformance). Then clients become frustrated and dissatisfied with both the resource and the case manager.

Working with Other Professionals

Clearly, effective service coordination depends to a certain extent on the case manager's relationship with other professionals. Both referrals and service monitoring are more easily achieved when the case manager has a relationship with the resource. The professionals on whom a case manager relies have a wide variety of cultural backgrounds, academic achievements, and job descriptions. Often barriers appear to service coordination that are rooted in turf issues, competition for clients, and concern about confidentiality. Communication, sometimes a challenge among those with different perspectives, is one way of addressing these barriers.

Good communication skills are critical when working with personnel from other agencies. These skills can be the deciding factor in making effective use of resources on the client's behalf. Here we present certain suggestions for enhancing communication with other helpers. First, avoid stereotyping other professionals. You may have encountered one nurse who was rude, but it is unreasonable to think that all nurses are that way. Second, don't hesitate to ask for clarification or a definition of terminology that you don't understand. It is better to ask than to pretend you know. Third, you can help others learn your own terminology by using it and explaining its meaning. Finally, be aware that other professionals may well have different styles of communication. For example, a clinical style may be more comfortable for psychiatrists. Other styles that have been identified are legal (of equal adversaries), political (of unequal adversaries), and pedagogical (teacher–student).

The Committee on Mental Health Services Group for the Advancement of Psychiatry (1981) lists 15 ways to achieve more effective communication between psychiatrists and decision makers outside the mental health field. Although this list was compiled some time ago, these 15 points are relevant for case managers, particularly as they engage in service coordination with a variety of other helping professionals.

BEFORE THE ENCOUNTER
1. Do your homework. Be as familiar as possible with the issues at hand and the answers you need to have.
2. Be conscious of the other person's circumstances—his values, his "language," his commitments—and respect them.
3. Be sensitive to the process of the interaction, especially the rules and practices of the specific situation in which you are involved.
4. Know your limits. Don't expect to be all things to all people. Use consultants or other resource persons who are familiar with the problem and the situation, and who can help you present your case.

DURING THE ENCOUNTER

5. Identify mutual concerns and common goals of both parties. Seek to develop an alliance for the purpose of solving the problem.
6. Define the issue and stick with it. Don't digress.
7. Listen. Keep cool, and don't argue. (At times, noncommunication may be the best communication).
8. Use simple language and concrete, familiar examples. Avoid technical jargon and difficult-to-understand abstract examples.
9. Don't bluff. Feel free to say "I don't know" and "I'll find out."
10. Keep your sense of humor, but direct it at yourself, not at your questioner.
11. Keep your narcissism and self-righteousness firmly in check.
12. Don't expect to win them all. At the point of impasse, back off and seek mediation.

AFTER THE ENCOUNTER

13. Remember that more encounters probably will occur in the future. Review what happened and learn from it.
14. Do not leave unresolved issues until the next time. (Some of the most fruitful exchanges can take place at times other than when an encounter is in progress.)
15. Informed contacts and meetings with decision makers should be encouraged. (Even if issues are not on the "agenda" or resolved, pathways for communication can be established by getting to know each other.) (pp. 162–163)

Using these 15 points can facilitate the work of case managers with other helping professionals. Being a skillful communicator and having good working relationships with other helping professionals enhances the case manager's role as a client advocate.

Advocacy

Advocacy is speaking on behalf of others, pleading their case or standing up for their rights. When case managers act as advocates for their clients, they are supporting, defending, or fighting for another person or group. Advocacy is also related to client empowerment and participation since case managers support clients' involvement in decisions about their own treatment and welfare. A caseworker at the Third Avenue Family Service Center in the Bronx describes a situation faced by AIDS patients that illustrates the need for advocacy.

It is called temporary housing, and as long as the client is living everybody is safe. If the client dies, then the family has to relocate. I have found that to be very unfair, because what happens to the children? There is no primary planning. . . . Nobody is working on this issue yet. (Carolyn Brown, personal communication, May 5, 1994)

Advocacy is very important for case managers. Often, clients are unable to articulate what they need, nor do they understand what choices are available to them. They may not have the necessary information or the skills needed to present their positions. In some situations, what they want is in direct conflict with people in authority. Clients may then be too intimidated to speak for themselves, or the people in power may refuse to consider client wishes. Terry is an example of such a client.

Terry is a 14-year-old who has just been diagnosed HIV positive. She is terrified of her medical condition and will not talk to anyone about it. She just sits and stares when the caseworker, her mother, or any other member of her family tries to address the subject. Her mother wants to send her away to a residential school, but Terry will not speak of this or any related issue.

In some agencies, the people in charge may not wish to hear client complaints or grievances. There may be no way for clients to appeal decisions or discuss methods of treatment they do not support.

Clients, by definition, are usually not in a position to act as advocates for themselves. Advocacy requires confidence, a feeling of control, and an understanding of the system. Most clients are in the human service delivery system because they do not have these characteristics. One caseworker at a school for the deaf tells how she remained an advocate even after her official responsibilities were completed.

In this particular instance, we had made more than one report to Human Services, and Human Services was already involved, but then closed the case. Although our reports were not the reports that reopened the case, the case was reopened and the children were removed. . . . What is unique about this particular case is that usually when they are removed, I don't have any further contact with the families . . . but I just knew that the kids were going to be returned, and I felt like I needed to maintain that relationship, because I would be making recommendations to the court for return. (Personal communication, October 27, 1993)

The following are some common client problems that case managers may consider appropriate for advocacy.

- Client has been denied services, or services have been limited.
- Client has expressed interest in one method of treatment but has been given another.
- Client's family has made decisions for him or her.
- Client has little information about the assessments gathered or the decisions made.
- Client has been denied services based on factors such as race, gender, or religion.

- Client has been given treatment contrary to his or her cultural norms.
- Client has been treated with disrespect by human service professionals.
- Client is caught in the middle of a conflict between two agencies or professionals.
- Client is being given unsafe or indifferent care.
- Client does not know what his or her rights are.
- Rules and regulations do not serve the client's needs.

Case managers are well positioned to provide advocacy for clients they serve. In the first place, case managers know their clients. They are most familiar with their skills, values, wishes, and the treatment they have received. A measure of trust has built up between the client and the case manager, so the case manager is likely to hear from the client if there has been unfair treatment. Case managers are also in contact with family members, friends of the client, and other professionals with whom the client is involved. The case manager therefore has immediate access to anyone who may be involved in any unfair practice. Finally, in the process of monitoring service delivery, the case manager may discover situations that warrant advocacy.

Advocacy is not an easy role to perform. Before an advocacy effort is begun, several tasks must be completed. First, the case manager gathers the facts of the situation and determines whether there is a legitimate grievance—whether making requests on behalf of the client would be justified. The case manager then decides whether advocacy is more appropriate than other methods, such as problem resolution or conflict resolution. Third, the case manager discusses the need for advocacy with the client and show willingness to speak for the client. He or she needs the client's approval before any advocacy takes place.

How to Be a Good Advocate

Many case managers do not like conflict, and they fear that, as advocates, they will not be successful in meeting the needs of their clients. Advocacy is complicated because case managers have loyalties divided among the agency, the supervisor, and the client. The case manager's personal values and beliefs may complicate the advocacy process. The following guidelines support the work of effective advocacy. These guidelines build on the knowledge, skills, and values presented in our previous discussion of the assessment, planning, and implementation phases of the case management process.

> *Know the environment in which the conflict takes place.* This is important when planning how to act as an advocate for the client. It is helpful to know the appropriate person to whom to make the appeal. Having a good referral network and a good relationship with many agencies is useful in understanding the environment.
>
> *Understand the needs of the client and of the other parties involved.* It is helpful if the advocate can understand why the other parties are in conflict with

the client or seem not to respect the client's wishes. If the advocate has this information before the appeal, strategies can be developed to meet some of the needs of all involved parties or, at the very least, to articulate the common ground.

Develop a clear plan for the client. The advocate must be clear about what the client needs and what the other parties must give or give up to meet those needs.

Use techniques of persuasion when appropriate. Many situations are suited to persuasion. Persuasion is used to support client needs while respecting the rights of other parties. It includes stating the problem clearly, presenting critical background information and facts, explaining why the situation needs to be changed, and detailing an acceptable solution.

Once there is agreement, it needs to be stated or written, and all parties need to agree.

Use more adversarial techniques when persuasion proves ineffective. Using techniques that challenge the system directly, the case manager makes a formal appeal or takes a case through legal channels in an effort to promote change on behalf of the client. In a grievance procedure or a legal challenge, the client has an opportunity to be heard, either verbally or in writing. Usually the client's case is heard by an impartial person or group, and a ruling is made. Another adversarial technique is to use the media to take the client's case to the public.

Terry's mother wants a solution to the problem right now. She is determined to send Terry away and has already made contact with three residential programs. She plans to move Terry somewhere as soon as possible. In reality, Terry's mother is frightened by what lies ahead for Terry and for the family. The only way she knows of to cope is to distance herself from the problem. The caseworker feels strongly that Terry should have a voice in the decision; she begins her advocacy work by trying to talk with Terry about what she wants.

Special circumstances are not necessary for case managers to assume the advocacy role. In the normal course of the job, a case manager encounters many opportunities to act as an advocate for the client. The case manager then integrates advocacy into other case management responsibilities. Each of the actions listed below represents the basic goals and values of case management discussed in Chapter 1. Not only do they characterize effective advocacy, but they are also good standards of practice.

- Provide quality services by involving a team of professionals and the client (or, when appropriate, a member of his or her family).
- Interact with the client to plan treatment that is congruent with his or her values and cultural orientation.

- Monitor the case and set goals and outcomes based on quality standards of professional care.
- Continually communicate with other professionals about issues relating to client rights.
- Plan treatment that takes into consideration the client's preferences, strengths, and limitations, and provide additional support if you anticipate that he or she will have difficulty.
- Speak for the client only when he or she gives permission.
- Educate the client about the agency's policies and procedures.
- Create an environment that facilitates decision making by the client.
- Educate the client about options in treatment and about the process of making the treatment plan, and discuss the barriers that may be encountered during the implementation phase.
- Work within the system to support, modify, and create policies. Know which professionals are involved in this work, and discuss your opinions with them. Volunteer for committees that consider policy issues.
- Become involved in the political process. Contact public policymakers.

Advocacy work is difficult, and the case manager must exercise good judgment in choosing when to attack barriers to the client's cause. At times, the need for advocacy may not be clear cut.

Martha Severn has just been asked by one of her clients not to report the $10 a week that she makes taking in laundry. Reporting this income would mean that the client will lose some of her scholarship aid for school. Instead, the client wants Martha to try to change the eligibility rule. Although changing the rule might seem to be a special favor for this particular client, Martha happens to believe that this policy is a good one.

In other instances the case manager may have to deal with competing interests—those of the agency and fellow staff members, as well as the client's.

James Dowling is a student intern at a mental hospital. He believes that one of his clients is being abused by a night technician. His client tells him of beatings during the night shift.

Ms. Wise is an elderly client who receives attendant care at home. The agency that coordinates Ms. Wise's care has a policy of keeping clients on home services for as long as possible. Cheryl Santana, a case manager at the agency, believes that clients remain much too long in home care. Cheryl makes recommendations for residential care, and the agency routinely rejects those recommendations.

Case managers must be aware of how their advocacy efforts are perceived by others. Many case managers believe that their first loyalty is to the institution or agency; others feel that they need to support the efforts of the team. Vigorous advocacy, at the expense of team camaraderie, may jeopardize the client's trust in the team. Another factor is that it is difficult to act as an advocate for certain clients: Some people *are* dishonest, greedy, or troublemakers. With such clients, the case manager must think clearly about the legitimacy of any client demands and approach the issues with fairness in mind.

 ## Teamwork

As stated earlier, working with others—**teamwork**—is a key component of the case management process. Professionals find themselves forming relationships with clients, families, and friends of clients, co-workers, other professionals, and other agencies. These relationships include working on teams, counseling families, and forming partnerships with other agencies, businesses, or governmental units and departments.

Treatment Teams

In meeting the needs of clients who have multiple problems—children, those with developmental disabilities, the elderly, and many other client populations—a coordinated team approach is necessary because since several professionals are involved. Sometimes referred to as the **treatment team,** this group of professionals meets to review client problems, evaluate information, and make recommendations about priorities, goals, and expected outcomes. Stacie Newberry, who works at a shelter for runaway girls, describes their staffing procedures.

> We have staffing every Monday. We discuss all the girls that are present, starting with those who have been assessed by the psychologist that day. We then go on to the other girls who are present to evaluate what progress they have made or are making. Because of our contract with the state, some of our girls can stay up to 30 days, which means we might staff them four times during their stay. (Personal communication, December 13, 1999)

Using a team to make decisions has numerous advantages. Working as a team, professionals can share responsibility for clients as well as the emotional burdens of working with clients who have difficult problems. When a group of professionals is involved, all the dimensions of the client's situation are more likely to be considered, and team members can get one another's viewpoints about the advantages and disadvantages of each decision made. In the team setting, helpers can share their expertise and knowledge as they focus on each client's unique set of needs and circumstances.

Types of Teams

One type of team used in case management is the **departmental team,** made up of a small number of professionals who have similar job responsibilities and support each other's work. Colleagues bring their most challenging cases to the team, and their co-workers help identify client problems and generate alternative approaches to treatment. Kim Ehlers, who works with adults with disabilities, talks about her work in the departmental staff meetings.

> We were really excited about moving clients into their own homes, but we were not really prepared. . . . Now we have a team meeting as soon as we decide somebody is going to move, and we figure out who will assume what responsibility. (Personal communication, October 14, 1995)

Departmental teams are particularly useful when decision making is difficult or client problems create a stressful situation for case managers. The departmental team shares information, offers opinions, and often makes group decisions about how to work with clients.

Thom Prassa, a case manager who works with children, describes the importance of teams: "I think the number one thing for me, why I have survived these last two and a half years, is really establishing relationships with your team, your other case managers" (personal communication, July 10, 1994).

Within the departmental team, the case manager can fill either of two roles: leader or participant. When assuming the leadership role, the case manager presents a case for review. He or she describes the client and the current status and summarizes conclusions and decisions made to date. The material is presented in such a way as to encourage feedback and dialogue with other colleagues. When the case manager is in the participant role, another colleague presents a case, and the case manager must listen, study the situation, and provide advice and counsel.

An **interdisciplinary team** is a different way to work with other professionals to provide services to the client. As the name suggests, this team includes professionals from various disciplines, each representing a service the client might receive. Often it includes the client or a member of his or her family. Early in the treatment of a client, the interdisciplinary team gathers and shares data, establishes goals and priorities, and develops a plan. In the later stages of treatment, the interdisciplinary team monitors the progress of the client, revises the plan, and often makes decisions about aftercare. The case manager is often the team leader, giving other team members a holistic view of the client, ensuring that the client or a member of the family is heard or representing their viewpoints, and conducting an assessment of client problems. In addition, the case manager is expected to monitor client progress between meetings, set the agenda for the meeting, give a summary of client progress, discuss next steps and any dilemmas, and help reconcile differences of opinion. At times, the team leader is placed in a difficult situation since each helper at the table has credentials in a specialized area of professional expertise. Many of them are accustomed to managing cases, being leaders, and making assessment and treatment decisions on their own.

One outcome of initial interdisciplinary team meetings is an organized, well-integrated plan designed to meet the goals that have been established to meet client needs. A good example is the Individual Family Service (IFS) Plan for Bill, Linda, and Jane Smith presented in Table 8.1. This plan supports Jane, who is developmentally delayed, and her parents. They are involved in Project Continuity, a project supported by the University of Nebraska Medical Center in Omaha, Nebraska. This program involves families as care managers for infants and toddlers with chronic illnesses and developmental disabilities. In this project, infants and toddlers are identified and receive comprehensive services to meet their own and the families' needs. These include the provision of services during hospitalization, assistance in transition from hospital to home, and follow-up when children return home. The service coordination functions include coordinating assessments and coordinating the IFS Plan. One component of the plan was the designation of responsibility to the professionals providing each of the services listed. Besides the parents, there is a team of four professionals involved in developing the comprehensive plan. Tasks are clearly stated, and an individual is assigned to oversee or perform each task (Jackson, Finkler, & Robinson, 1992).

Once the treatment plan begins, interdisciplinary team meetings are scheduled to help monitor the client's progress. Each of the professionals involved presents a progress report, and together they make a decision about how to proceed. The case manager often meets with each of these professionals one-on-one before the meeting. In crisis situations, he or she may have to revise the plan without team approval. Interaction with other professionals can sometimes be difficult if they have not provided the services for which they are responsible or if there is some question about the quality of service delivered. The case manager is most often not their supervisor, so when such issues arise, he or she manages by persuasion and collaboration.

Teams with Families and Friends

Because case management is a viable model for serving clients with long-term, complex needs, families and friends are often an important part of the team. When working with clients such as the elderly, people with limited mental capacities, children and youth, the mentally ill, and other populations who depend on family and friends to help make decisions, provide care, or both, the case manager must recognize that the caregivers expect to be involved in the planning for services. There is also a trend to include families in treatment of pregnant teens, single mothers, parolees, immigrants, and others who were traditionally given individual treatment. Input and participation by families and friends is viewed as important at each phase of the case management process.

Working with families and friends can be rewarding and challenging. The benefits include expanding the network of support and care for the client, adding another perspective on the environment and needs of the client, and engaging

....

TABLE 8.1   INDIVIDUAL FAMILY SERVICE PLAN FOR JANE SMITH

Bill and Linda Smith	Parents
Joanie Dinsmore	Nurse Specialist
Penny Lees	Child Life
Jean Tomasek	Social Work
Barb Jackson	Education

Identified outcomes	Plan	Person responsible
1. The Smiths will receive developmental support for Jane.	a. Developmental assessment will be completed prior to discharge.	B Jackson
	b. Referral will be made to Bellevue Public Schools.	Smiths
	c. Parents will be given suggestions for appropriate developmental intervention.	B Jackson
	d. Inpatient intervention times will be arranged with the parents.	P Lees
2. The Smiths will be directed to resources providing family support, to use at their discretion.	a. Parents will be given information on Nebraska Respite, Pilot Parents, and Family Friends.	J Dinsmore
	b. Ms. Jones from Bellevue Public Schools will be contacted to identify other military families of children with special needs who may serve as a resource.	Smiths
3. The Smiths will have the opportunity to review financial resources for medical care and equipment.	Arrange for consultation with Jean Tomasek, MSW.	J Dinsmore J Tomasek
4. The Smiths will receive supportive information on intervention with their daughter, Jane.	a. Child Life will consult with the Smiths about information and visitation issues for Jane.	P Lees
	b. The Smiths will be provided with a list of available resource books for siblings.	P Lees J Dinsmore
5. Jane learns to communicate with others. She uses several ways to signal for more, e.g., smiles, vocalizes, increases movement, quiets.	It is important to set up a variety of situations where you can play simple games like rocking, talking to, or pat-a-cake with Jane. Observe to see if Jane begins to signal for more. These signals might be in the form of movement, a smile, or quieting, which is her way of saying that she wants more. If any indication is given, you immediately respond and play the game some more. Also, see if she begins to anticipate any familiar games.	Parents Pediatric nursing staff

TABLE 8.1   *(continued)*

Identified outcomes	Plan	Person responsible
6. Jane anticipates familiar routines and games.		B Jackson P Lees J Bell
7. Jane sits balanced with little support from an adult.	Seat Jane on the mat between your legs facing away from you. Dangle toys in front of her to keep her attention and gradually reduce your support to see if she can sit for brief periods. If she falls back, your body will block the fall.	Smiths
8. Jane pivots on her stomach to retrieve toys. Jane bears her weight on one hand/arm while she reaches for and plays with a toy.	Present toys to Jane as she is on her stomach, slightly out of reach on her right or left side. See if she will move to attempt to retrieve it. If necessary, physically guide Jane's hips and shoulders in a pivoting movement to reach a toy. When Jane is on her stomach, offer a toy and see if she will shift her weight in order to reach for it.	Smiths

SOURCE: From *Family-Centered Service Coordination for the 90s*, by J. Dinsmore and B. Jackson, Appendix C. Copyright © University of Nebraska Medical Center. Reprinted with permission.

those in the immediate environment as part of the solution. Not all family and friends facilitate the teamwork process: Family members may not agree with each other, do not support the client's commitment to change, and try to sabotage the case management process. Most case managers would still rather have family members on the team even if they are not totally supportive.

One example of broad inclusion of family is the Memory and Aging Project Satellite (MAPS) of the Washington University Alzheimer's Disease Research Center. This program was developed to meet the multiple needs of the underserved elderly with cognitive impairments. Many of these clients had no formal diagnosis of dementia. Members of the home-based services interview both clients and caregivers living in and outside of the home. One difficult problem encountered in this program is that often multiple caregivers do not provide the same advice when assessing needs and making recommendations (Edwards, Baum, & Meisel, 1999).

Benefits of Teams

Working on a team is an exciting experience for most case managers. They welcome the creative thinking and the support that comes from a collaborative effort.

However, building an effective team requires the efforts of all the members, especially the case manager. In most interdisciplinary teams, the case manager has the responsibility for leading the team and developing an atmosphere conducive to good **collaboration**. A positive atmosphere in which a team functions well occurs when there is a common goal. In the case of teams involved with case management, the goal is the successful development and implementation of a plan that meets client needs. Team members must have respect and trust for the others. Respect is important because teams share in decisions that can radically alter clients' lives. Mutual respect is especially important in interdisciplinary teams since each member is relied on to bring knowledge and skills in a particular area of expertise. In the case of departmental meetings, the participating colleagues continue to work side by side, so it's important to maintain respect (Prpic, 2001).

For the case manager who is involved with teams, several aspects of teamwork can directly improve the services provided and the work environment of the professionals involved. First, the clients receive services from several professionals working together. The greater the sum of expertise, creativity, and problem-solving skill applied, the more effective the planning and delivery of services will be. Each professional can perform his or her responsibilities better because of having participated in the process of setting goals and priorities as well as planning. Because of the team, the professionals also have a better sense of the client as a whole person and a clearer conception of how their own particular treatment is integrated into the larger plan.

Teamwork also enhances members' skills in making decisions and solving problems creatively. Not only do they have opportunities to practice those skills, they can also learn from the other professionals. The environment fostered by teamwork is valuable to any helping professional, but especially to the case manager who is coordinating services. Good communication skills are developed in a team atmosphere; members learn to listen well and to speak to the group when appropriate (Lauer, 1996). The opportunity to share responsibility—to ask for assistance, to volunteer or give it, and to receive it—helps all team members by increasing their sense of community and reducing isolation (Lauer, 1996). At the Jewish Home and Hospital for the Aged, one case manager used team meetings to discuss difficulties, and it brought positive results.

> We would talk about difficulties, because it is very difficult to deal with some of the residents. I mean, they are not the most pleasant sometimes, and if they are sick and dying and demanding on top of that, it becomes really a tremendous burden on staff . . . so we created an open forum where people could say "I can't take it." (Personal communication, May 3, 1994)

Teamwork is central to performing the case management role of service coordination. Linking clients to services, monitoring client progress, and communicating with other professionals are important components of effective service delivery.

Chapter Summary

Service coordination is a key to the case management process. Because it is the case manager's responsibility to locate resources, make arrangements for clients to use them, and monitor client use and progress, coordinating services is a critical phase of case management. The helper and the client agree on a plan of service first and then determine what services can be provided by the case manager and what services need to be referred. It is important for the client to participate during resource selection so that his or her values and preferences and unique circumstances are considered. Because the case manager often does not provide all the services needed, referral is an important component of service coordination. Effective referrals take into account matching available services with client needs. Readiness of the client to be referred, appropriateness of the referral, and readiness of the new agency to receive the client are all important factors in making a successful referral.

Once a referral is made, it is the case manager's responsibility to monitor the services provided and the client's progress. This includes a periodic review of the services, a note of changing conditions either on the part of the client or the services being provided, and an evaluation of client progress. Evaluation of client progress is an ongoing responsibility and includes investigating the status of problems, client satisfaction with the referring agency and staff, the status of outcomes, and determination to continue or close the case. There are guidelines that can facilitate working with others on this assessment. Case managers are encouraged to use good communication skills, to know their own limits and the limits of other helpers and agencies, to listen well, and to encourage dialogue about issues of disagreement. Referral and monitoring among helping professionals, at its best, occurs in an atmosphere of understanding and mutual trust.

Advocacy, a key element in service coordination, focuses on providing the best services available to the client. Advocacy often requires speaking and acting on behalf of clients when they are not able to speak for themselves. Sometimes organizations and individuals must respond more fairly to the clients they serve; clients may be denied services; services have been limited; or the client may be caught in a conflict between two professionals or two agencies. Whatever the situation, part of service coordination is helping clients negotiate the system, working to ensure that they receive the most effective services available.

Meeting client needs often requires working in teams. Teamwork can take place within departments, across departments and agencies, and with clients and their families. The benefits of using teams include the ability to serve clients with multiple needs and to increase the number of resources available.

Chapter Review

◆ *Key Terms* ....

Resource selection   Teamwork
Referral   Treatment team
Mobilizer   Departmental team
Monitoring services   Interdisciplinary team
Advocacy   Collaboration

◆ *Reviewing the Chapter* ....

1. Name the three activities of service coordination.
2. What are the benefits of service coordination?
3. How would you use a systematic resource selection process when making a referral?
4. Under what circumstances does a case manager refer a client?
5. Discuss three reasons why referrals fail.
6. What are the steps in a successful referral?
7. Describe how the roles of broker and mobilizer apply to the monitoring of services.
8. What are the three components of monitoring services?
9. State some guidelines for working with other professionals.
10. Define advocacy.
11. Name some common client problems that reflect the need for advocacy.
12. What are some guidelines for good advocacy?
13. How does advocacy relate to the goals and values of case management?
14. What different types of teams are encountered in case management?
15. What are the benefits of teams?

◆ *Questions for Discussion* ....

1. After reading this chapter, what evidence can you give that coordinating services is a critical component of the case management process?
2. Why do you think advocacy is an important responsibility for a case manager?
3. Suppose you were a leader of a case management team. How would you conduct your first meeting?

References

Brill, N. (1998). *Working with people: The helping process.* White Plains, NY: Longman.
Committee on Mental Health Services Group for the Advancement of Psychiatry. (1981). *Interfaces: A communications case book for mental health decision makers.* San Francisco: Jossey-Bass.

Edwards, D. F., Baum, C. M., & Meisel, M. (1999). Home-based multidisciplinary diagnosis and treatment of inner-city elderly with dementia. *The Gerontologist, 39*(4), 483–488.

Jackson, B., Finkler, D., & Robinson, C. (1992). A case management system for infants with chronic illnesses and developmental disabilities. *Children's Health Care, 21*(4), 224–232.

Lauer, M. (1996). *Team building: Training the wheels together.* Unpublished paper, University of Tennessee, Knoxville.

Prpic, J. K. (2001). Teamwork. cleo.eng.monash.edu.au/teaching/subjects/learning/bridging/teamwork.html

Working Within the Organizational Context

We have different goals for outreach and different levels of success. For some people we meet, success is getting them into a shelter and getting them off the streets. There are people who don't want to get off the streets and don't want a place to stay. Maybe they just want us to help them get through the winter or get their food, or their warm clothes. We try to get clients what they want. The clients are not always trying to get the same thing.
— Rick Langley, Youth in Need, St. Louis, MO, personal communication, December 13, 1999

Most of the Healthy Start Program is built on collaboration or partnerships. A lot of the services in this program are provided by other agencies since the state Healthy Start program only gives us $400,000 for four years.
— Norma Cervantes, Roosevelt High School, Los Angeles, CA, personal communication, March 24, 1998

The program began to meet the basic needs (of homeless)—no questions asked. About two years ago, we realized there was no accountability. We never knew what happened to the person. There are other outreach efforts in the city who are doing the same thing. We felt we were duplicating their services. So we have changed our program and added a bike patrol.
— Deno Fahbre, St. Patrick's Center, St. Louis, MO, personal communication, December 14, 1999

The case management process takes place in the context of an agency, and client services often involve more than one organization. Effective case managers understand the organizations to which they belong. This requires mastery of three key concepts: organizational structure, agency resources, and improving services. The preceding quotes illustrate the importance of these three concepts in case management.

The organizational structure of an agency reflects the agency's mission, goals, and policies. Knowledge of these aspects allows the case manager to understand his or her actual job responsibilities and working environment. According to Rick Langley at Youth in Need, the goal for providing outreach to the homeless is to respond to client needs as expressed by the client, not just to find them housing. This mission of the agency is reflected by the wide range of services given the clients, including meals, clothing, bathing facilities, and recreational opportunities. All these services reflect a strong commitment to letting clients determine the services they want.

An agency's budget constraints are also important. Even though the Healthy Start budget is limited, the program at Roosevelt High School offers a wide variety of services to students and their families because they work with other agencies in Los Angeles. With these partnerships, the Healthy Start Program is able to provide parenting classes, child care, welfare case management, mental health care, and many other services. Without these partnerships, the services would be limited to basic health care.

Case management work also involves trying to improve services, with an emphasis on meeting client needs. This often means evaluating outcomes or periodically reviewing services and cost effectiveness. The Mobile Outreach Program at St. Patrick's Center changed its service delivery to avoid the duplication of services, to establish measurable outcomes for the program, and to increase their effectiveness in meeting the needs of the street people in St. Louis.

During our discussion of organizational factors, you will meet Carlotta Sanchez, who works for the Sexual Assault Crisis Center in a city of 400,000. She has just begun her responsibilities as a case manager for the agency. Along with Carlotta, you will learn about the organizational structure of the Sexual Assault Crisis Center, the budgeting process and her role in it, the agency's informal structure, and the organizational climate in which the center's work is performed. You will also read about her participation in agency programs designed to improve the quality of services.

For each section of the chapter, you should be able to accomplish the following objectives.

UNDERSTANDING THE ORGANIZATIONAL STRUCTURE
• Name three ways that knowledge of organizational structures benefits a case manager.
• List some sources of information about an organization.
• Differentiate between formal and informal structure.

MANAGING RESOURCES
• Trace the planning and budgeting process of an organization.
• Identify other ways in which the case manager is concerned with resource allocation.
• Name the components of a budget.
• List sources of revenue.

IMPROVING SERVICES
• Identify four processes for improving the quality of services.

Understanding the Organizational Structure

To be an effective case manager, one must understand the organization and its structure. This knowledge is helpful in three ways. First, it gives the case manager a better understanding of his or her job responsibilities and how they relate to the goals of the agency and the specific objectives of the unit. Second, case managers can help meet client goals more easily if they use agency procedures correctly. Finally, once case managers understand the organizational climate, their work can fit appropriately into the context of the agency. They know what is expected of them, how much autonomy they have, and who can help with the difficult situations they face. Case managers face obstacles in any setting; knowledge of the particular working environment helps to identify barriers and develop strategies to cope with them.

The Organization's Plan

Several documents may shed light on the way an agency is structured. One is the agency's mission statement. Other important statements are its goals, objectives, and policies and procedures (Ellis & Hartley, 1999; Gillies, 1994; Lewis & Packard, 2000; Morgan & Hiltner, 1992; Porter-O'Grady, 1994). As these documents are examined, it is helpful to ask the following questions.

> What is the mission statement of this organization?
> What values are reflected in this mission statement?
> How do these values influence the work of case management?
> What are the goals and objectives of the agency?
> How do these relate to the work of the case manager?
> What are the written policies and procedures of the organization?
> What impact do these policies and procedures have on the case manager?

A **mission statement** is a summary of the guiding principles of the agency. It usually states the broad goals of the organization very concisely and describes the populations that will benefit from the work of the organization. It may also specify the values that guide all decisions, the agency structure, sources of funding, agency priorities, and the work of the staff (Ellis & Hartley, 1999; Lewis & Packard, 2000). The mission statement for the National Institute of Mental Health, a federal agency responsible for research in mental health and mental illness, follows. Notice that it is succinct, has a clear focus on research, and targets the public as the recipient of services.

> The mission of the National Institute of Mental Health (NIMH) is to diminish the burden of mental illness through research. This public health mandate demands that we harness powerful scientific tools to achieve better understanding, treatment and, eventually prevention of mental illness. (www.nimh.nih.gov/about/index.cfm)

Each organization has a set of policies and procedures, some of which pertain to the behavior of all employees. Such documents are often very specific, describing procedures in great detail. Many reflect a standard of practice that is determined by the agency, as well as legal requirements established by the federal, state, and local governments. Also included are standards of practice established by a professional code of ethics or by a professional accrediting body. The Mitchell Area Adjustment Training Center in Parkston, South Dakota has a policy on the composition of the review board that approves applicants for services: It must include the executive director, a service coordinator, a nurse, and one representative from the other three units of service. At many agencies, the speed with which clients move through the system is regulated by stated policies and procedures. Pima Health System has such a time line. "Three days after they [clients] are determined eligible, they are enrolled with us. . . . Within 10 days of enrollment with us, a case manager has to do a face-to-face visit with the client and develop a care plan with the client and with the client's family members" (personal communication, October 7, 1994).

Once the case manager has a clear picture of the agency's direction, guidelines, and rules, it is important to know the job description, the job's relationship to the work of the department or unit, and the agency's expectations. Case managers should ask the following questions of their supervisors or the people who are responsible for hiring new employees (Ellis & Hartley, 1999; Gillies, 1994).

Where can I find a written description of my job as a case manager?
What are the goals of my department or unit?
How does case management fit into those goals and objectives?
Does the department or unit have its own policies and procedures?
How do these policies and procedures relate to my job as a case manager?

Rarely does the description of a position accurately reflect the actual work, but it does serve as a guideline. The **job description** does define the work for which the case manager is held accountable, but it often changes with reorganization, economic pressures, and the changing needs of the client population. Judith Slater describes how her job as case manager has evolved. "When I started doing case management, there was just me and my boss. . . . I covered every specialty area of the hospital. . . . It was pretty much referral by doctors. With the shift in economics and with Medicare–Medicaid guidelines, we then had to screen every Medicare patient. Initially it was referral, and then it was screening out certain types of diagnoses and discharge planning" (personal communication, October 14, 1995).

Structure of the Organization

Another way to understand the structure of the organization or agency is to discover the relationships among people and departments. The following questions are helpful (Ellis & Hartley, 1999; Gillies, 1994; Morgan & Hiltner, 1992).

Who has authority or control over others?
Who is accountable to those in authority?
What are the supervisory patterns within the agency?
What is the communication flow within the agency?

In applying this information to your work as case manager, consider how the relationships define authority, accountability, and communication. The following questions will be helpful.

> To whom do I report?
> Who evaluates my work?
> Does anyone report to me?
> Do I evaluate anyone?
> How do I get information?
> To whom do I pass information?

Two terms important in understanding the structure of the organization are *authority* and *accountability.* **Authority** refers to controlling resources and action (Ellis & Hartley, 1999). The lines of authority in the organization become clearer when departments are examined. Responsibility to others in the organization for what you do and how you use resources is **accountability** (Ellis & Hartley, 1999). Simply stated, a major component of authority is holding individuals accountable for the jobs they perform. Each person in authority also has to be accountable for that which he or she controls.

The **chain of command** of an agency stretches from the position with the most authority to the one with the least. Policy information is often passed along the chain of command, from the top down.

At Helen Ross McNabb Mental Health Center, Jana Berry Morgan and Paula Hudson explain how the chain of command is used to change an implementation plan for a client.

> If the person that they are working with didn't have day treatment . . . and the case manager felt like that was a necessity for them, then the case manager can go through the client's primary clinician first. Then referrals are passed to day treatment. Then the day treatment coordinator would do an assessment of the referral and determine if the client was an appropriate candidate for day treatment. (Personal communication, April 27, 1995)

Many layers of professionals may be involved in making this decision, each person having authority for part of the decision and accountable for that part.

A document that is helpful when determining the chain of command is the **organizational chart.** (See Figure 9.1.) This chart is a symbolic representation of the lines of authority and accountability, as well as the information flow within an organization. An organizational chart usually consists of an arrangement of boxes that represent offices, departments, and perhaps individuals.

The boxes at the top of the chart represent the positions with the most authority, supervisory experience, and control of resources. Boxes that are connected with solid lines represent departments, offices, or individuals for which there is a formal communication pattern. For example, any communication from the Executive Director would go to the four directors of the agency.

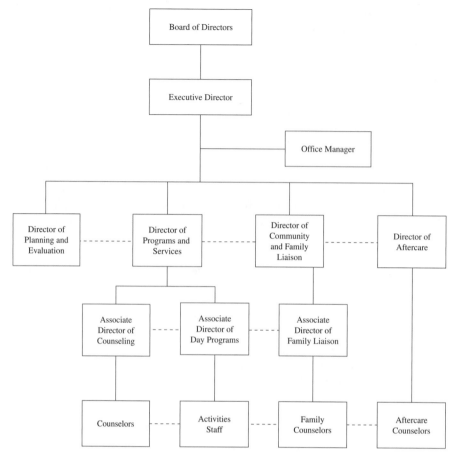

Figure 9.1   Organizational chart

Boxes that are connected with vertical solid lines represent lines of authority. The Executive Director is responsible for the entire organization and is accountable to the Board of Directors. The Associate Director of Counseling and the Associate Director of Day Programs are supervised by the Director of Programs and Services. These two associate directors in turn supervise the counselors and the activities staff. Boxes connected by dotted lines represent departments, offices, or individuals who communicate with one another but do not have supervisory authority or control of resources. For example, all the directors communicate with one another as they plan and implement the work of their departments.

Final authority in many human service agencies and organizations rests with the board of directors. The primary responsibility of the board is the financial health of the organization. Under normal conditions, board members do not work directly with supervisors, staff, or clients, but they do interact with the highest-ranking individual in the organization. The board focuses much of its attention on the budget (Russo, 1993). Members of the board analyze statistics such as the characteristics of clients served, the staff-client ratio, the sources of funding, and the relationship between expenditures and outcomes. The board is also involved with public relations, helping the agency establish its mission and goals, and working for a positive public image (Morgan & Hiltner, 1992; Russo, 1993).

Let's see how Carlotta Sanchez is adjusting to the organizational structure of her agency.

Carlotta has been in her job as a case manager for six months now. It has been quite a learning experience for her. She believes that she was hired so quickly in part because she had worked as an intern with a women's shelter during her senior year. Carlotta is not really sure how many women she has been able to help in the past six months, but she knows that she has been learning a lot. She is finally able to work with clients and families without constantly asking for help from her friend Sally, who is also a case manager. Sally has been in her job for three years and was a volunteer before that, so she understands the work of the agency well.

During the first month, Carlotta had difficulty in understanding how to prioritize her work. She was coming in an hour early and staying two hours later than the regular staff just to complete the work she thought she should be doing. During those first months, everyone seemed to think they knew what she should be doing, but Carlotta herself was confused about her role. Her supervisor, Ms. Ludens, who oversees the work of the five case managers, is rarely available to talk with Carlotta about the job, although she is pleasant and supportive. At the end of that first month, Sally pulled Carlotta aside and asked if she needed any help.

Carlotta asked Sally if she could help her plan her work schedule. Carlotta also called her college advisor to talk about the difficulties she was experiencing. Both Sally and her teacher asked Carlotta several questions that she could not answer, ones that focused on the agency's policies and procedures. Carlotta barely remembered her orientation, which had consisted of two days of lectures on agency policy and rudimentary training in working with the target clients (primarily women who had been victims of sexual assault). During that orientation, she also got information about the legal system. At that time, everything was so new that she didn't know what questions to ask.

Carlotta asked whether she could attend orientation again the next time a new case manager was hired. She reread the mission statement and examined the goals and objectives. Some of them looked really familiar, but there were others that she did not remember, and she was not sure the agency ever addressed them in its programming. According to the mission statement and the goals and objectives, the Sexual Assault Crisis Center is dedicated to prevention and education. Carlotta had never contributed to those goals, and she did not see such activities going on in any other unit. Everyone seemed to be too busy working with clients who had already been assaulted or who were in danger.

After reading the policies and procedures, she was confused by discrepancies between what they said and what her colleagues did. She noted one particular serious violation of policy. All case managers are to follow a time line after receiving a referral. Intake is to occur within 48 hours and is followed by a staffing, but Carlotta has never been able to meet that deadline. She talked to Sally about her questions and concerns. Sally just looked at her and told her to ask the supervisor; Sally would not say another word. Carlotta thought that was a strange reaction, so she decided to ask Manuel, another case manager; he had never heard of the policy that she was questioning. She decided that Ms. Ludens must be the person to turn to.

Carlotta scheduled a meeting with Ms. Ludens. She came up with a list of questions that concerned her, including her job in relation to the mission, the time line that she had not been able to follow, and her need for more supervision. In preparing for the meeting, she found an organizational chart that showed a Director of Case Management, reporting to the Executive Director.

Carlotta felt that the meeting with Ms. Ludens went well, although she still had unanswered questions afterward. Ms. Ludens very carefully reviewed the job description for the case manager and prioritized the responsibilities of case managers. Carlotta discovered that Ms. Ludens expected her to conduct intake interviews, coordinate a team that included all the professionals working with a client, work with the family, and provide long-term yearly followup with clients. She did not expect her to participate in educational or preventive activities. At times it seemed that Ms. Ludens did not understand the job of the case manager. Ms. Ludens provided a curt answer to the question about the mysterious Director of Case Management: "We no longer have that position." She gave Carlotta an equally short response when asked about the timing of intake and staffing: "That is not my policy."

Although she was still unclear about the conflict between policy and practice, Carlotta left the meeting knowing a bit more about the organization.

An organization's planning documents, the written job description for a case manager, and the chain of command (as illustrated by an organizational chart) give a picture of the formal structure. However, as in Carlotta's experience, the formal organization does not give any case manager all the necessary information about the agency. Each agency also has an informal structure that influences the work that is done and how it is accomplished.

The Informal Structure

The **informal structure** of an agency supports many of its functions. The agency's informal structure also helps meet the needs of supervisors, staff, and case managers. The case manager must distinguish between the formal structure and the informal structure. According to Russo (1993), the major differences between the two are as follows.

FORMAL	INFORMAL
Who supervises whom	Who advises whom on what to do
What the job description says	What the individual in the job actually does
What the formal lines of communication are	Who communicates with whom
What the policy is	What actually happens

An informal structure develops within every organization as a way of meeting the needs that are unmet by the formal structure. It enables workers to use relationships and alternative ways of getting the job done. This structure facilitates the work of the case manager and helps meet social and professional needs; co-workers celebrate successes and support each other in times of trouble. They form a network of support and cooperation based on mutual respect and long-term relationships. As discussed earlier, such a network helps the case manager locate services and monitor the progress of clients.

The informal structure of any agency has existed long before the case manager assumes responsibilities, and he or she also becomes a part of that structure. For example, through informal channels, co-workers give the new case manager advice on defining client problems, addressing issues of client motivation, finding services, and using agency policy to help meet client needs. Often, co-workers provide support to each other when one of them encounters difficult clients or stressful situations, and they rejoice together when their work goes particularly well. At St. Patrick's Mobile Outreach Program, staff always go out in pairs on bikes or in the van. "We try to get a feel for the situation by talking with the street person. We always leave them with information about our programs" (personal communication, December 14, 1999). This is a particularly helpful strategy when working with clients outside the office.

The informal structure also involves an alternative pattern of communication flow (Ellis & Hartley, 1999). More individuals in the organization thus receive useful information outside of channels. Having access to information supports the decision-making process. For example, case managers may hear about changes in policy or procedure or budget cutbacks before they happen. Advance notice gives them time to adjust their work for the coming changes or seek to modify them, serving as advocates for clients' interests.

Although the informal structure contributes to the strength and the success of an organization, it can also contribute to difficulties. For example, the informal structure tends to represent tradition, preservation of the status quo, and resistance to change (Ellis & Hartley, 1999; Gillies, 1994). This hinders case managers from advocating change to help meet the needs of their clients. In addition, communication within the informal structure may be unreliable or even hurtful. Such informal information must be considered carefully but not regarded as gospel (Ellis & Hartley, 1999). Several roles, such as advocate, broker, service organizer, and expediter, depend on the case manager's status as a representative of the agency. Passing on unreliable or inaccurate information or making decisions based on such information may damage his or her professional credibility.

Since there is no official map or diagram of this informal structure, case managers must construct their own. Russo (1993) suggests that case managers make diagrams of the interaction patterns that they observe. The following questions about one's own case management activities provide insight into one's interaction with the informal structure.

Which colleagues provide me with information?
Which colleagues do I help?
Who is my supervisor?
Whom do I supervise?

There can be many combinations of answers to these questions. Drawing a chart based on the answers and then comparing it with the organizational chart can help illuminate the differences.

Roz Jaffe, who works in a day care setting for the elderly, describes an informal way of sharing information about clients with other staff, from the moment the clients arrive at the agency. "The drivers let us know if there is a problem. Or someone tells me they are not feeling well or there may be a problem with their home care worker. I will go to the nurse or the social worker to make sure they are aware of the problem or potential problem" (personal communication, May 3, 1994). This everyday exchange of information is very different from the formal weekly staff meetings, care plan meetings, or the charting that is completed at the end of each day.

Another helpful way of looking at the organization is to analyze the organizational climate.

The Organizational Climate

The conditions of the work environment that affect how people experience their work is the **organizational climate.** Such a climate is difficult to describe, for it is based on the values, attitudes, and feelings of people in the work setting. Case managers need to understand the climate so as to clarify their work expectations and their degree of autonomy. Ultimately, the organizational climate influences how they perform their jobs and how they relate to their clients. Two major factors influence the organizational climate: (1) policies and procedures and (2) supervision (Ellis & Hartley, 1999).

As discussed earlier, each organization has a set of policies and procedures that guides the behavior of employees. These can give one of three signals to employees (McGregor, 1985a, 1985b; Ouchi, 1981):

1. We trust you will do a good job if we supervise you closely.
2. We trust you to do a good job because you want to do a good job.
3. We trust you to do a good job because you are loyal to the agency, which is like a family.

Case managers within the first, highly supervised environment have job responsibilities that are very clear. The policies and procedures are narrowly defined and well articulated. In the second, less structured environment, policies and procedures allow more autonomous decision making, and case managers have less rigid job descriptions and responsibilities. In the third, "family" framework, an agency invests much time and energy in its case managers and expects loyalty from them in the form of long-term commitment and a focus on excellence (Ellis & Hartley, 1999; Gillies, 1994; Morgan & Hiltner, 1992).

How a case manager is supervised also contributes to the organizational climate, and the supervision is not always consistent with the agency's stated policies and procedures (Ellis & Hartley, 1999). The supervisor can establish a tone that encourages independent thinking and activity, or he or she can narrow the range of responsibility and decision making. Even within an organization that has strict policies, the supervisor can foster a less rigid climate. For example, if the supervisor encourages autonomy, case managers may have the freedom to determine their own schedules, make final recommendations for client treatment, and establish agreements with other organizations, all without prior approval (Ellis & Hartley, 1999). Thom Prassa and Mark Alexander describe how the supervision structure of their agency affects their work: "We have in our office . . . divided responsibilities into a Coordinator, two Supervisors, seven Case Manager II's and 15 Case Manager I's. . . . It is a very loosely governed office. . . . We are given a lot of latitude to develop our own professional style, because it is a new program and because each region has its own peculiarities" (personal communication, July 10, 1994).

Let's now go back to the organization in which Carlotta works.

After Carlotta met with Ms. Ludens, she decided that she really needed to work more closely with her colleagues if she was going to improve her performance and help her clients better. Sally had been very helpful, and Carlotta decided that she would also work more closely with Charlotte Jones, the legal liaison for the center. Ms. Jones has been with the agency for ten years and has a wealth of knowledge about the law, relations with the police and the community, and the welfare of clients. In staffings, everyone listens when she makes recommendations. When she seeks Ms. Jones's advice, Carlotta noticed that many of her colleagues were doing the same. It also seemed that Ms. Jones was willing to answer any questions that Carlotta asked.

Carlotta broached the subject of the 48-hour turnaround from intake to staffing. She honestly felt that she was not doing her job when she had to schedule staffings after the 48-hour period. Ms. Jones smiled when she heard Carlotta's question, and she answered it straightaway: "So you have noticed. Let me tell you why we have that policy. Our Board of Directors feel that 48 hours is a reasonable time to begin to work on a case. They also believe that we have not really begun our work until the staffing takes place. We have explained many times why intake is often a longer process than 48 hours, but they refuse to change the policy. Our last director lost her job over that battle. Ms. Ludens just refuses to acknowledge the policy. We have all decided to do the best we can. We know the policy; we try to follow it. Sometimes it is just not possible."

Carlotta also asked Ms. Jones about the Director of Case Management position that she had seen in the organization chart. Ms. Jones shook her head before she spoke. "That was another battle we lost with the board. Our budget was cut two years ago, and the board recommended that we reduce our indirect personnel costs. They wanted us to put most of our resources in direct services to clients. With the difficulty of this work, the turnover for case managers is high, and training our new case managers is critical. Your questions are evidence of that need for supervision."

Several months later, Carlotta was meeting Sally for dinner after work. Carlotta had actually left work at the conclusion of office hours that day. She told Sally that she was beginning to understand her work better, and she said, "I am even feeling more at home in the agency."

Providing case management within the context of an agency is an exciting and challenging activity for Carlotta and other case managers. Another challenge is managing resources. The next section discusses how resource allocation influences the work of the case manager.

 ## Managing Resources

Understanding the basic concepts of budgeting is important for a case manager. Much of the planning within the agency relates directly to the allocation of resources. Resource allocation determines the number of staff, the degree of operating support, the number of clients served, and the programs and services available for clients. Allocation of resources even covers the amount of time case managers spend on specific tasks.

Since costs are rising, with a simultaneous increase in the number of clients who have complex needs, there is an imperative to spend every dollar well. Allocating resources wisely requires an understanding of budgets and their applications to case management work.

What Exactly Is a Budget?

Gillies (1994) describes a **budget** as "a numerical expression of an agency's expected income and planned expenditures for a period of time" (p. 85). According to Gross and Jablonsky (1979), "A budget is a 'plan of action.' It represents the organization's blueprint for the coming months, or years, expressed in monetary terms" (p. 359). Budgeting has two purposes: planning and controlling. Sometimes the planning and controlling functions of a budget have opposing goals; planning represents the creative and expansive approach to human service delivery, whereas control is restrictive and measured. However, both are important facets of the budgeting process and have great relevance to the case management process.

Experts agree that planning and budgeting should be linked and that planning should guide the budgeting process (Gillies, 1994; Whalen, 1991). A human service agency must be clear about goals, objectives, and priorities before it begins budget preparations (Ellis & Hartley, 1999; Gillies, 1994; Lewis & Packard, 2000; Morgan & Hiltner, 1992). A two-step process links planning and budgeting. The first step is to set up three categories: goals and objectives, planned activities, and the costs of those activities (Ellis & Hartley, 1999). For example, an initial part of the budgeting process for the Mitchell Area Training Adjustment Center was to submit a request to the state for "eight new staff to take the individuals [clients] out into the community at least once a week" (personal communication, October 4, 1994). Planners at the center believe that without eight new instructors, the parents of the clients will complain that their sons and daughters are not receiving quality care.

The second step in the budget process considers the goals and objectives and the resources available to provide the services. Revision of the budget then occurs within the parameters of the available resources. The result of the second step is a budget with revised costs to reflect the goals, objectives, programs, and services as adjusted to the funds actually available. This process allows planning to determine the budget (Gillies, 1994; Lewis & Packard, 2000; Morgan & Hiltner, 1992). Once the Mitchell Area Training Adjustment Center receives its allocation from the state, it will complete the second step of the planning and budgeting process—revising the goals for clients on the basis of the state's response to the request for eight new staff.

The case manager's work in the planning and budgeting process is determined by the agency's policies and procedures. There is now a management trend toward participatory management and budget responsibility at the unit level, so case managers may be asked to provide input into the process (Gillies, 1994; Whalen, 1991). Because case managers have knowledge of client needs, departmental goals and objectives, and new professional trends in research and service delivery, they can offer helpful recommendations for priorities and resource allocation.

Many case managers are involved directly in budgeting for their individual clients. Some have responsibility for managing an account of services for clients; they generate and monitor a budget for each one of them. This includes defining needs, setting priorities, procuring services, and paying for services rendered. Usually this begins with a fixed allocation of funds and a list of services that may be purchased for clients who have a specified set of needs. Case managers help set priorities and determine how best to use the funds. A case manager at the Pima Health System has these responsibilities: "There are expenditure limits and there are cost-effectiveness studies done for each client. The cost of home- and community-based services cannot exceed 80% of what it would cost to provide nursing home care to the client" (personal communication, October 7, 1994).

Resource allocation extends to how case managers use their time. Time is a valuable resource of the case manager, to be used for various activities, such as intake, planning, coordinating resources, monitoring client progress, advocacy, providing counseling, educating, interdisciplinary teamwork, building networks, and completing paperwork. Time management of each day is a resource decision, and the use of time should be linked to outcomes.

Another purpose of a budget is to establish control of expenditures (Ellis & Hartley, 1999; Gillies, 1994; Morgan & Hiltner, 1992; Porter-O'Grady, 1994; Whalen, 1991). The budget provides a baseline from which to judge and project obligations. If an organization adheres to the budget during the year, the budget controls spending. It is important to maintain a balanced budget, in which expenses do not exceed expenditures; this is one standard of good fiscal management.

The control function of the budget has a relationship to case management work in terms of how programs are implemented and how needs are met. Planning takes into account the resources available and any restrictions on their use. There may be policies governing what services clients are eligible to receive (Ellis & Hartley, 1999; Gillies, 1994; Lewis & Packard, 2000). For example, the Family Counseling Center in Tucson has multiple funding sources: Social Services block grants, Title III, the Older Americans Act, and the county funds. The agency's policy is to ask clients to pay for some of the services if they can. Limited resources may also restrict the number of clients that can be served. Sometimes, planners must develop eligibility criteria that consider the prevailing limitations on resources.

Features of a Budget

A budget has several components: projected expenditures, actual expenditures, and balance. (See Table 9.1.) *Projected expenditures* are cost estimates made at the beginning of the budget cycle; they designate the money that has been set aside for expenses in a given category. *Actual expenditures* represent the money that has been spent in a given category to date. At any time in the budget year, it is clear what has been spent in each category. The *balance* is the amount allocated to a given category minus the amount already spent.

In Table 9.1, expenditure categories are listed in the left column.

Salaries are money paid to employees, either for providing direct services to clients or for indirect services, such as administration and supervision.

TABLE 9.1   BUDGET FORMAT FOR A HUMAN SERVICE AGENCY

Category of expenditures	Projected expenditures	Actual expenditures	Balance
Salaries			
Supplies			
Equipment			
Travel			
Training			
Communications			
TOTAL			

Supplies are necessities for maintaining the activities of the agency. Usually, such items have a short life or are used within a year. Examples are paper, pens, gasoline, and essentials for repair and cleaning.

Equipment includes machines that are bought for a specific purpose, such as computers, a copy machine, and recreation equipment. These are usually one-time expenditures.

Travel represents the cost of transporting personnel and clients. Some agencies cover clients' travel expenses to and from sites where services are delivered. Case managers may make home visits or transport clients to other agencies. The travel budget is also used for purposes of professional development.

Training costs often include payment to consultants who conduct staff training for professional development, as well as the cost of certification classes, books, and professional journals.

Communications includes expenditures for using telephones, sending faxes, mailing, and any other expenditures incurred in transmitting information.

Total designates the whole amount that has been committed or spent.

Often the budget is simplified by using just three categories: salaries, operating, and capital. The *salaries* category is the same as that for the budget presented in Table 9.1. *Operating* includes supplies, most equipment, travel, training, and communications. The *capital* budget includes major equipment and building projects.

Sources of Revenue

Funding for a human service agency or organization determines if an agency is public or not-for-profit (voluntary). **Public** or governmental **agencies** exist by public mandate and receive funding from one or more of the following sources: the federal, state, regional, county, and municipal government. State departments of human services or children's services are examples. **Not-for-profit** agencies or organizations are funded by individual contributions, fundraisers, foundation grants, and corporate donations. They are also governed by an elected board of directors. St. Patrick's Center, which you read about at the beginning of the chapter is an example of a not-for-profit agency. Today the distinction between public and not-for-profit agencies is not always clear-cut because not-for-profit agencies are increasingly providing services for public agencies on a contractual basis (Federico, 1980).

Recently, a third category—**for-profit agencies**, has proliferated. Reasons for their increase are the limited resources for voluntary agencies, reduced governmental funding, and changing political and economic times (Alle-Corliss & Alle-Corliss, 1998). Two functions guide service delivery by a for-profit organization: (1) providing a service and (2) making money. Managed care organizations, health maintenance organizations, and private corporations are examples.

Whether public, not-for-profit, or for-profit, agencies and organizations often receive funding from multiple sources. Four main revenue streams fund human service programs: (1) federal, state, and local government-sponsored funding, (2) grants and contracts, (3) fees, and (4) private giving (Ellis & Hartley, 1999; Gillies, 1994; Lewis & Packard, 2000).

Federal, state, and local government funds   Government funds are available for programs and services. As in the case of Medicare and Medicaid, they may be direct reimbursements for client care. In the past, this was considered stable long-term funding. In today's climate of rising costs, balanced government budgets, and the diminishing role of government in human services, such funding is less secure. There is a movement to allocate financial support to local levels of government in order to increase accountability and to develop programs closer to where the problems occur.

Grants and contracts   Many agencies write proposals for grants and try to develop **contracts** to secure funding from both government agencies and private corporations. Staff members are sometimes assigned to write proposals to acquire grants and contracts. Sponsoring agencies often establish priorities and set restrictions on resource allocation. If human service agencies and their target populations are to benefit from such external funding, the priorities of the granting or contracting body must be similar to those of the agency (Ellis & Hartley, 1999; Lewis & Packard, 2000).

Fees   Revenue from **fees** is generated in two ways. First, many clients have insurance that covers the costs of the services they receive. The insurance is managed through a third party, and the payment is made according to the policies of the insurance agreement. In the managed care environment, there may be restrictions on the range or quantity of services available (Hicks, Stallmeyer, & Coleman, 1993). Fees for services are also collected from individuals who choose not to use their insurance, who have exceeded the amount the insurance will pay, or who are ineligible for government or insurance support. In such cases, fees are often assessed on a sliding scale: The higher the person's income, the higher the fees; the lower the income, the lower the fees.

Private giving   Private giving is becoming an important source of revenue. Human service agencies solicit individuals, businesses, and corporations for donations of money, equipment, and professional expertise. The current trend is to establish relationships with such donors in the hope that funding can be stabilized by a long-term partnership.

How an agency is funded influences the case management process. Government funding and insurance reimbursements have direct effects on the services provided. Any change in funding has the potential to expand, restrict, or alter services. Many programs have changed over time as funding patterns have shifted. For example, Angela LaRue, a parole officer, describes changes in an adjustment counseling program for parolees. "We used to have an adjustment counseling therapist. The therapist did this for free for a while. Then we got funding. . . . Then he did it for free when we lost the funding" (personal communication, October 18, 1995).

Case managers may be involved in grant writing or may serve as consultants to those who are making applications; working on grant-writing projects gives them an opportunity to speak for the needs of clients. In addition, case managers may be involved in fundraising. For many years, private fundraising was the sole responsibility of the board of directors and the executive director. As private giving becomes a more important part of agency revenue, case managers can expect to increase their participation in fundraising activities. The case managers at Casita Maria solicit private donations to enhance the services they provide. In one Casita Maria program, clients receive an allowance to buy a bed and a kitchen set for a new apartment. With private donations, caseworkers are able to get their clients additional furniture. Yolanda Vega, Director of Agency Services, shared one client's appreciation: "She was more than thrilled, because she has got her apartment in order" (personal communication, May 4, 1994).

Carlotta Sanchez, the new case manager at the Sexual Assault Crisis Center, has little experience with budgets. Let's see how she begins to learn about the resource allocation process at the agency.

To Carlotta, it seemed that the more she talked with her colleagues, the more she discovered the importance of the Board of Directors. She knew that Ms. Ludens spent quite a lot of time with board members, and during the week of a board meeting, the whole staff was busy guessing what changes the meeting might bring for the agency's functioning. Ms. Jones had also told Carlotta that funding was very tight for the coming year. Carlotta had discovered that

*although the budget year for the agency began January 1 and ended December
31, budget preparations began very early in August.*

---

*Carlotta still did not have a supervisor, and she had no way of getting key in-
formation about the agency unless Ms. Ludens spoke at the large staff meeting
or sent an "all-staff" memo. Sometimes, Carlotta heard news from the other
case managers or Ms. Jones. By the middle of the summer, several of her colleagues were expressing worry about the budget for the coming year. Several
funding sources were cutting back, and the staff had no way of guessing how
the agency would be affected. The state government had experienced a short-
fall for the second year in a row and was talking about reducing support to social services. United Way had increased the list of agencies it supports and
therefore might decrease its funding to the center. Donations from the public
sector were also down from the previous year.*

---

*In September, Ms. Ludens called the entire staff together and asked them to
prepare the budget for the coming year. They were to consider three scenarios:
a budget increase of 2%, a budget decrease of 2%, and a budget decrease of 4%.
She established a budget team and asked them to describe what impact the increase scenario and the two decrease scenarios would have on programming
and staffing. She also asked the budget team to consult with everyone in the
agency. Carlotta was chosen to represent the case managers on the team.*

---

*Carlotta learned a lot during the two months she worked with the budget
team. They talked about priorities, the cost of programs, the target population,
and how to set and measure expected outcomes. Carlotta was amazed at how
her own responsibilities would change in each scenario. It was a stressful period, and involving the entire staff in the budget planning took a lot of time and
effort. The group had the three scenarios ready for Ms. Ludens by the deadline,
and she prepared to present them to the board at its meeting the following
week. The whole staff was holding its breath; the actual budget total would not
be available until final word about state funding arrived.*

One other effort related to organization influences the roles and responsibilities of case managers in the delivery of human services. The next section discusses how agencies involve their case managers in the process of improving
services.

 ## Improving Services

In recent years, there has been an increasing emphasis on providing quality services. This has resulted in the use of four processes: linking outcomes to cost, conducting utilization review, planning quality-assurance programs, and promoting
continuous improvement. Each process addresses the issue of quality in a different way.

What Is Quality?

Quality is a term that is very difficult to define. Persig (2000, p. 163) expresses the dilemma this way.

> Quality . . . you know what it is, yet you don't know what it is. But that's self-contradictory. But some things are better than others, that is they have more quality. But when you try to say what quality is, apart from the things that have it, it all goes poof! There's nothing to talk about. But if you can't say what quality is, how do you know what it is, or how do you know it even exists?

In the fields of health and mental health care, *quality* is understood to mean that the delivery system meets its objectives (with a particular emphasis on client needs) in relation to four variables: the limitations of the methodology used, the client's status, the environment, and the resources available (Brook & Lohr, 1985; Davis & Meier, 2001; Donabedian, 1983; Savitz, 1992; Steffen, 1988). Defining quality and measuring it are only one component in providing quality services. There must then be an identification of any problems and a corresponding change in service delivery (Savitz, 1992). This section presents two approaches to defining and measuring quality: linking objectives to outcomes and conducting utilization reviews. We also discuss two approaches to improving quality: conducting quality-assurance programs and implementing continuous improvement programs.

One way to identify the quality of services provided is to evaluate the outcomes of programs in relation to the resources spent (Davis & Meier, 2001; Gillies, 1994). To do this, programs and the necessary resources for implementation must be weighed against expenditures. Outcomes such as the number of clients served, the programs delivered, and client progress are matched to actual expenditures. To perform such an evaluation, Feldman (1973, p. 44) suggests providing answers to the following questions.

What are the objectives of this organization?
What programs are available to move toward these objectives?
What will each of these programs cost in human and financial terms?

The results of this analysis are helpful to a case manager, for one of the most frustrating elements of case management is the difficulty of determining one's own effectiveness and efficiency (Gillies, 1994; Lewis & Packard, 2000; Morgan & Hiltner, 1992). If case managers can begin to link their activities to the objectives of the organization (progress made by the client) and to the cost of the service delivered (combining the costs of direct service to the client with any support services), they begin to see the relationship between their work and the benefits to their clients.

One type of quality analysis has emerged from the managed care environment: utilization review, which has been developed to oversee service provision and monitor costs. It is a response to accusations that managed care substitutes cost effectiveness for quality care. In the area of mental health, managed care has also been criticized for underfunding necessary treatments (Knesper, Belcher, & Cross, 1987; Savitz, 1992; Sharfstein, Krizay, & Muszynski, 1988; Wyszewianski, Wheeler, & Donabedian, 1982). Let's examine utilization review and its impact on the case management process.

Conducting a Utilization Review

A utilization review is conducted by peers within the human service delivery system. Two approaches are used during utilization review: pretreatment review and second-opinion mandates (Corcoran & Gingerich, 1994; Tischler, 1990). In **pretreatment review,** a managed care professional must review and approve a treatment plan before service delivery. **Second-opinion mandates** apply to certain diagnoses: Consultation with a second professional must take place before a treatment plan can be approved. Gillies (1994) states that utilization review is designed to determine the proper treatment for each of the client's problems or conditions, the appropriate course of the treatment process as a whole, and how the submitted plan must be revised.

Utilization review is often applied because the treatment is provided on a prepayment basis. In other words, since limited funds are allotted for each client, it is critical to assess client needs and the cost of relevant services before providing such services. The responsibility of people who perform utilization review is to provide for the treatment measures that represent the best use of resources in light of the anticipated outcomes.

In many managed care PPOs, health care and mental health care are provided to employees of a corporation, or to members of a designated group at discounted rates, in return for a high volume of clients. For cost-effective management, the diagnosis and treatment regimen must be shortened (Feldman & Fitzpatrick, 1992; Gillies, 1994; Mullahy, 1998). The ultimate goal is to spend the least amount of time and provide the lowest-cost treatment that restores the client to good health or mental health (Feldman & Fitzpatrick, 1992; Gillies, 1994; Mullahy, 1998). Another goal is to keep professionals out of legal difficulties. To meet these goals, another process has developed, one that emphasizes quality care. The process of quality assurance documents that the care provided has met quality standards (Porter-O'Grady, 1994).

Planning Quality Assurance Programs

The focus of **quality assurance** programs in the human services is on developing standards of client care (Davis & Meier, 2001; Gardner, 1992; Mullahy, 1998; Porter-O'Grady, 1994; Savitz, 1992). Relevant to the work of the case manager are two quality-assurance measures: the standard treatment plan and the client satisfaction survey.

An agency develops a standard treatment plan to ensure common practices in intake, assessment, goal setting, planning, implementation, and evaluation of outcomes. Agency practice must establish standards of care for common assessments within the scope of the agency's activities. The work of the case manager and others must then be compared with the standards established. Thus, each case manager provides the same set of services for each client. When services do differ, reasons must be provided. For the quality-assurance process to be of real value, the case manager must be able to link assessment with implementation and with positive outcomes.

Client satisfaction is another component of quality-assurance programs. As a dominant criterion for evaluating their work, it is viewed both positively and negatively by professionals. Some would question the client's competence to judge the quality of services (Savitz, 1992). On the other hand, there is evidence that clients who are pleased with their treatment are more likely to follow the plan and to be straightforward in reporting whether their goals have been met (Steffen, 1988). Since case management aims to empower clients and increase the extent of their responsibility for their own lives, client satisfaction can be considered a key element in providing quality service.

Figure 9.2 shows how client satisfaction is measured by one agency that uses service coordinators to manage cases. Investigating satisfaction is difficult; it requires an atmosphere in which clients believe that someone wants to hear what they have to say. It is also helpful to hear from clients who terminate services prematurely or whose treatment proves unsuccessful. Their identification of problems in service delivery is beneficial to agencies that value quality assurance, for this feedback can serve as the basis for needed improvements. Unfortunately, such clients are sometimes difficult to reach.

The final approach to improving the quality of services is **continuous improvement** (also known as total quality management). This movement began in industry but is fast becoming an accepted part of quality assurance in the human service delivery system (Feldman & Fitzpatrick, 1992; Hicks, Stallmeyer, & Coleman, 1993). Supervisors and case managers each have a role in the continuous improvement process. Supervisors are charged with providing leadership that encourages all staff to participate in improving service delivery; case managers are expected to provide quality work and to collaborate with others in a spirit of cooperation and teamwork.

Edward Deming (1986), a leader in the total quality movement, suggests several guidelines to be observed if work toward quality in an organization is to succeed. First, note that *all* processes and procedures can be improved, so each component of the process is subject to scrutiny and possible change. Another guideline of Deming's is that customer needs (in this case, those of clients) must be considered. A third component of continuous improvement is the requirement of teamwork across units and departments to solve problems and improve outcomes. Clearly, this goal is consistent with the commitment of the case manager to work effectively with colleagues.

If an agency is participating in a total quality program, the case manager may be asked to conduct peer reviews or join a quality circle. In a *peer review* system, professionals establish a process for evaluating one another. Another feature of continuous improvement is the *quality circle,* in which those who serve on case management teams work together to solve problems.

1. Date contacted by case manager _____

2. Number of meetings with case manager _____

3. How would you rate the frequency of contact with the case manager?

   Adequate _____   Too much _____   Too little _____

4. Were the case manager's goals and intervention consistent with your needs?

   Definitely _____   Somewhat _____   Not at all _____

5. Please describe your diagnosis. _____

   _____

6. Has the status of your difficulties changed since you had your first contact with our agency?

   _____

   _____

   Please describe: _____

   _____

   _____

   _____

7. Check the appropriate column for the following.

	EXTREMELY HELPFUL	SOMEWHAT HELPFUL	HELP IN THIS AREA WASN'T NECESSARY
Effectiveness of our services in dealing with your situation	_____	_____	_____
Understanding your problems and intervention	_____	_____	_____

Figure 9.2   **Quality-assurance questionnaire**

Obtaining information
from your case manager
to share with you (e.g.,
treatment alternatives,
instructions to be
followed)   _____   _____   _____

Assisting with
arrangements for
necessary services   _____   _____   _____

Offering support,
encouragement, or
assistance in coping
with your difficulties
or those of a family
member   _____   _____   _____

8. Were there other ways in which the case manager was helpful? Please describe:

   _____

   _____

   _____

9. How would you rate the work of your case manager?

   Excellent _____   Good _____   Fair _____   Poor _____

10. In terms of effectiveness, our agency intervention came

    At the right time _____   Too early _____   Too late _____

11. Were there other areas of assistance that could have been helpful? Please list:

    _____

    _____

12. How would you like to see our services changed? _____

    _____

    (*continues*)

Figure 9.2 (*Continued*)

---

13. Would you recommend our services to others in a similar situation?

Yes _____   No _____

Signature: _____   Date: _____
(Optional)

Figure 9.2   *(Continued)*

SOURCE: Adapted from *The Case Manager's Handbook,* by C. Mullahy, pp. 260–262. Copyright © 1998 Aspen Publishers, Inc. Adapted with permission from the author.

Another area of focus for the continuous improvement approach is the management of time (mentioned previously in the section on managing resources). Managing time well means knowing how time spent is related to priorities and expected outcomes. Good time-management techniques for the case manager are discussed in Chapter 11.

Let's revisit Carlotta as she becomes involved in her agency's efforts to institute continuous improvement.

Once she had participated in the budget team, Carlotta's confidence increased, and she felt that she was able to add something of value to the agency. She began to look around for other committees that she might join. She felt that she was too new to the organization to join the Personnel Committee, and she was not very interested in fundraising and development activities (dealing with large crowds and planning events were not much fun for her). However, she was interested in the total quality management (TQM) program that was beginning to be applied to the agency's volunteer training. The TQM group included two volunteers, the volunteer coordinator, a board member, a client, and a member of the Mental Health Association. The group wanted a member who worked directly with clients, and they invited Carlotta.

In the beginning, Carlotta thought that she had never been to so many meetings. They had training on six consecutive Saturdays just to become familiar with the principles of TQM. Carlotta's father worked for a chemical company in a neighboring state, and he had told her about his terrible experiences with quality teams. He thought that they were a waste of time and gave workers false hopes that things would get better. Carlotta was more optimistic, and she really enjoyed meeting and working with people she normally had little contact with. In the first sessions, the group learned about TQM and the progress it had fostered in other organizations. By the third Saturday, the group had started focusing on the volunteer program.

First, they had to establish the agency's mission, goals, priorities, stakeholders, and customers. Carlotta thought that the agency only had one customer—the client—but she learned differently. After much discussion, it was decided that the agency had many customers, among them the board, the citizens of the county, the local government, and the police in the city and the county. The list was much longer, but the point for Carlotta was that more people depended on her agency than she had ever realized.

The first six Saturdays were only the beginning of the meetings, and the group spent a lot of time outlining what happens in the agency, exactly how the work flows, and the role of volunteers in that work. They learned about the training and supervision of volunteers, their actual work experiences, how clients responded to working with them, and how the volunteers themselves felt about their work. They sent out a survey and then interviewed clients and volunteers about their experiences to determine what they valued about their experiences with the agency and how satisfied they were. Carlotta was not sure exactly how this work would turn out, but it gave her a different perspective, and she enjoyed working with these new colleagues. She came to be confident that all her work would help provide better conditions for volunteers as well as clients.

Chapter Summary

Case managers work within an organizational context and must understand the essentials of agency structure, how to manage resources, and how to improve services. Understanding these fundamentals enables them to use the organization more effectively to meet client needs. They can then see more clearly how their work is integrated into the mission of the agency.

Many case managers are also expected to manage resources. A thorough understanding of the budget process and its effects on the resources available for clients helps case managers serve as effective advocates for their clients' interests. This includes knowing the sources of funding, distinguishing between public (governmental) and not-for-profit (voluntary) agencies, and understanding the impact of for-profit agencies on service delivery today.

Case managers are naturally committed to improving services to meet the changing times, resources, and client needs. Providing quality services occurs in four ways: linking outcomes to cost, conducting utilization reviews, planning quality-assurance programs, and promoting continuous improvement.

Chapter Review

◆ Key Terms

Mission statement	*Public agencies*
Job description	*Not-for-profit agencies*
Authority	*For-profit agencies*
Accountability	*Contracts*
Chain of command	*Fees*
Organizational chart	*Pretreatment review*
Informal structure	*Second-opinion mandates*
Organizational climate	*Quality assurance*
Budget	*Continuous improvement*

◆ Reviewing the Chapter

1. Why is it important for a case manager to understand organizational structure?
2. List sources of information about an organization and describe the type of information each provides.
3. How can a case manager find out about relationships within an organization?
4. Describe the relationship between authority and accountability.
5. What is the purpose of the informal structure? How does it operate, and how is it different from the formal organizational structure?
6. How does the organizational climate affect the work of the case manager?
7. Name the two steps that link planning and budgeting.

8. What role do resources play in preparing a budget?
9. Describe how budgeting for clients takes place.
10. In what other ways does resource allocation affect case management?
11. Name the components of a budget.
12. Design a budget for yourself for one week, using the categories discussed in this chapter.
13. How does an agency get money to support its work?
14. Distinguish between public, not-for-profit, and for-profit agencies.
15. Describe the two ways in which fees are generated.
16. How does the agency's funding affect case management?
17. How do linking objectives to outcomes and conducting utilization reviews relate to defining and measuring quality?
18. How do quality-assurance programs and continuous improvement programs contribute to better services?
19. Describe pretreatment reviews and second-opinion mandates and their roles in utilization review.
20. How does the use of standard treatment plans affect case management?
21. Discuss client satisfaction as an element in quality case management.

◆ *Questions for Discussion* ....

1. Why do you think it is important to understand an organization's informal structure?
2. Can you provide a rationale for the statement, "Case managers should understand the process of managing resources"?
3. What do you think will happen as case managers learn more about how to improve services?
4. Do you think it is a good idea to involve clients in the process of improving services? Why or why not?

References

Alle-Corliss, L. & Alle-Corliss, R. (1998). *Human service agencies: An orientation to fieldwork*. Pacific Grove, CA: Brooks/Cole.

Brook, R. H., & Lohr, K. (1985). Efficacy, effectiveness, variation and quality: Boundary crossing research. *Medical Care, 23,* 710–722.

Corcoran, K., & Gingerich, W. J. (1994). Practice evaluation in the context of managed care: Case-recording methods for quality assurance reviews. *Research in Social Work Practice, 4*(3), 326–337.

Davis, S. R., & Meier, S. T. (2001). *The elements of managed care: A guide for helping professionals*. Belmont, CA: Wadsworth.

Deming, W. E. (1986). *Out of crisis*. Cambridge, MA: MIT Press.

Donabedian, A. (1983). The quality of care in a health maintenance organization: A personal view. *Inquiry, 20,* 218–222.

Ellis, J. R., & Hartley, C. L. (1999). *Managing and coordinating nursing care* (3rd ed.). Philadelphia: Lippincott.

Federico, R. C. (1980). *The social welfare institution: An introduction.* Lexington, MA: D. C. Heath.

Feldman, J. L., & Fitzpatrick, R. J. (Eds.). (1992). *Managed mental health care.* Washington, DC: American Psychiatric Press.

Feldman, S. (1973). Budgeting and behavior. In S. Feldman (Ed.), *The administration of mental health services.* Springfield, IL: Charles C. Thomas.

Gardner, D. (1992). Measures of quality. In M. Johnson (Ed.), *The delivery of quality health care* (pp. 42–58). Chicago: Mosby.

Gillies, D. (1994). *Nursing management.* Philadelphia: Saunders.

Gross, M. J., & Jablonsky, S. F. (1979). *Principles of accounting and financial reporting for nonprofit organizations.* New York: Wiley.

Hicks, L., Stallmeyer, J., & Coleman, J. R. (1993). *Role of the nurse in managed care.* Washington, DC: American Nurses Publishing.

Knesper, D. J., Belcher, B. E., & Cross, J. G. (1987). Preliminary production functions describing change in status. *Medical Care, 25,* 222–237.

Lewis, J. A., & Packard, S. (2000). *Management of human service programs* (3rd ed.). Pacific Grove, CA: Brooks/Cole.

McGregor, D. (1985a). Theory X: The traditional view of direction and control. In *The human side of enterprise* (pp. 33–43). New York: McGraw-Hill.

McGregor, D. (1985b). Theory Y: The integration of individual and organizational goals. In *The human side of enterprise* (pp. 45–57). New York: McGraw-Hill.

Morgan, E. E., & Hiltner, J. (1992). *Managing aging and human service agencies.* New York: Springer.

Mullahy, C. M. (1998). *The case manager's handbook.* Huntington, NY: Aspen.

Ouchi, W. G. (1981). *Theory Z: How American business can meet the Japanese challenge.* Reading, MA: Addison-Wesley.

Persig, R. (2000). *Zen and the art of motorcycle maintenance.* New York: Bantam Books.

Porter-O'Grady, T. (Ed.). (1994). *The nurse manager's problem solver.* St. Louis, MO: Mosby.

Russo, J. R. (1993). *Serving and surviving as a human service worker* (2nd ed.). Pacific Grove, CA: Brooks/Cole.

Savitz, S. A. (1992). Measuring quality of care and quality maintenance. In J. L. Feldman & R. J. Fitzpatrick (Eds.), *Managed mental health care* (pp. 143–158). Washington, DC: American Psychiatric Press.

Sharfstein, S. S., Krizay, J., & Muszynski, I. L. (1988). Defining and pricing psychiatric care "products." *Hospital Community Psychiatry, 39,* 372–375.

Steffen, G. E. (1988). Quality medical care. *Journal of the American Medical Association, 260,* 56–61.

Tischler, G. L. (1990). Utilization management of mental health services by private third parties. *American Journal of Psychiatry, 147,* 967–973.

Whalen, E. L. (1991). *Responsibility center budgeting.* Bloomington: University of Indiana Press.

Wyszewianski, L., Wheeler, J. R., & Donabedian, A. (1982). Market-oriented cost containment strategies and quality of care. *Health and Society, 60,* 518–550.

Chapter Ten

Ethical and Legal Issues

*C*ase managers wear lots of hats. Advocating for the client when needed . . . and one of the things in our system is the gatekeeper role . . . But the case manager really works for the client and for the HMO too . . . so you can't just be a client advocate. You have to determine what services are truly needed and appropriate for the client.
   —Alan Tiano, Pima Health System, Tucson, AZ, personal communication, October 7, 1994

*O*ne of the most difficult issues is how not to help when the client doesn't want it. How to back off when the client needs space. That clients have the right to fail. That is hard, because we see potential. We want them to get better. We know that if they [would] just take this medication, they would get better, but they don't want to.
   —Jan Cabrera, Intensive Case Management, Los Angeles, March 24, 1998

*N*ow legally the home attendant is not supposed to give medication to a person. The person can say, "Gee, it's time, and here it is," but they are not legally supposed to do it. So she doesn't give him medicine at night, sometimes he could use something for sleeping and I didn't blame her for not doing it, and you know there is nobody else to do it.
   —Rosalyn Baum, Jewish Home and Hospital for the Aged, Bronx, NY, personal communication, May 3, 1994

I have a client whose family thinks that she should not be on the medicine. Her family tells her that we are bad for her and . . . she listens to her family . . . and then she sees what they are telling her may not be in her best interest.
   —Jana Berry Morgan, Helen Ross McNabb Center, Inc., Knoxville, TN, personal communication, April 27, 1995

Effective human service delivery often requires a delicate balance of consideration to the client, the agency for which the case manager works, laws and regulations, court rulings, and professional codes of ethics. These conflicting interests can create crises that require the case manager to make difficult choices. The chapter-opening quotations reflect some of the tensions that case managers face. Alan Tiano describes the conflict resulting from case managers' dual responsibility to their clients and to the agency. Advocacy means working for the client's interest, and gatekeeping entails upholding the rules, regulations, and restrictions of the agency, which are not always best for the client.

Jan Cabrera speaks of a different type of dilemma, involving the mandate to grant client autonomy whenever possible. Case managers may see clients choosing alternatives that are not in their best interests. Sometimes, it is very difficult to let clients make these choices. Roz Baum talks about the difficulties involved in providing services according to legal or ethical guidelines. Clients often violate rules just so they can maintain their stability and have their needs met. Professionals always have to assess what they think and how they will behave in these situations.

Working with families is often integrated into the case management process. Sometimes families can enrich and support the helping process; sometimes they can cause difficulties. Jana Berry Morgan describes a time when family pressure threatened one key component of the case management plan.

In situations such as those described above, as well as many others, finding the appropriate resolution is difficult and challenging. Case managers must constantly ask themselves certain questions: What is in the client's best interest? What is the right choice ethically? Am I operating within the guidelines of the agency that employs me? Case managers look to codes of ethics, the law, and agency policies and procedures to guide their practice. Mediating disagreements among family members and clients, working with potentially violent individuals, honoring client preferences, maintaining confidentiality, and upholding complicated rules and regulations are among the thorny issues with which case managers grapple. Several pressures increase the challenges case managers face. Clients are becoming more aware of their right to make decisions about their own care, and families are becoming more involved in their relatives' care. New technological, psychological, and economic interventions are continually being developed. Dealing with finite resources, case managers must control costs and allocate resources equitably.

For each section of the chapter, you should be able to accomplish the objectives listed.

FAMILY DISAGREEMENTS
- Define the "daughter from California" syndrome.
- List guidelines to follow to encourage positive participation of families.

WORKING WITH POTENTIALLY VIOLENT CLIENTS
- Describe why violence is becoming more prevalent in modern society.
- Apply the steps in addressing issues of violence in the workplace to a specific case management situation.

CONFIDENTIALITY
- List reasons why the issue of confidentiality is so difficult.
- Define ways that managed care and technology have affected confidentiality.
- Describe guidelines to follow when discussing confidentiality with clients.
- Describe the ways that technology affects client confidentiality.

DUTY TO WARN
- Define duty to warn.
- Demonstrate how the case manager works with a team on issues involving the *duty to warn*.

AUTONOMY
- Describe the difficulties that arise with regard to granting client preferences.
- Explain how guidelines can help a case manager who faces issues of autonomy.
- Describe how case managers can support autonomous end-of-life decisions.

BREAKING THE RULES
- List the sources of rules and regulations.
- Tell how to decide when to break a rule.

 ## Family Disagreements

Working with clients and their families is an important part of case management. Often, families can provide support to clients that will help them meet their goals. According to the American Counseling Association, counselors should "recognize that families are usually important in client's lives and strive to enlist family understanding and involvement as a positive resource, when appropriate" (American Counseling Association, 2001). One of the most difficult ethical areas for the case manager involves end of life issues. One such issue is the **"daughter from California" syndrome**—family disagreements over the care of an incompetent client. This syndrome is named for the case of Mrs. M., an elderly woman whose two daughters disagreed vehemently about her care (Molloy, Clarnette, Braun, Eisemann, & Sneiderman, 1991).

> Mrs. M. was a woman 83 years of age with a five-year history of Alzheimer's disease. She had been cared for by her 60-year-old daughter and placed in a hospital when the daughter could no longer handle her. Because of medical complications, the mother was unable to walk, incontinent, and demented. The daughter recommended a "do not resuscitate" (DNR) order. She insisted that her mother be given medical treatment only for comfort and pain. The daughter was sure that this measure coincided with her mother's wishes, although there was no written directive from the mother.
>
> Complications arose when a second daughter, who had not seen her mother in years, arrived from California. This daughter was outraged with the DNR order and demanded that her mother receive treatment to continue her life for as long as possible. The daughter from California was angry with her sibling and denied that her mother was in a serious state of decline. She dominated treatment team discussions and threatened to sue the entire team. The team worked with this daughter to help her with her denial, guilt, and anger. The sister who had previously provided the primary care gave in to the California sister's wishes, and aggressive medical treatment for the mother resumed. The California sister went home, and the DNR order (with treatment only for comfort and pain) was put into effect once more. The mother died two weeks later, and the daughter from California did not return.

The legal question in such situations is this: Who can make the decision to terminate life-prolonging measures when the patient is incompetent and has left no clear directions? In 1990, the United States Supreme Court, in a 5 to 4 decision, upheld a ruling from the Missouri Supreme Court that "it is constitutionally permissible for a state to require that life-prolonging measures for an incompetent patient remain in place unless the patient had left behind a clear directive excluding such measures in the event of a future state of incompetency" (Molloy et al., 1991, p. 397). The state rule is not mandatory, but when an agreement cannot be reached, the court becomes the advocate for the patient and makes the decision.

This case is one example of the complex decision making that must occur when the wishes of the client, the family, and helping professionals are not in agreement. The situation becomes more complicated when the goals of the organization or the ethical standards of the professionals are also being challenged. Molloy and colleagues (1991) propose the following hierarchy of decision making to help professionals proceed when the patient is incompetent and the family cannot agree on the treatment.

1. If the patient while competent had signed an advance directive, then his or her wishes must prevail, even over the contrary views of family members.

2. If the patient has left no directive, it is necessary to turn to the family. An in-family discussion of treatment options is called for, although, as a general guideline, actual decision-making authority should be allotted in accordance with the following prioritized list: spouse, adult children, siblings, other family members. (The term *spouse* includes the patient's significant other.)

3. If there is irresolvable in-family conflict regarding case management of the patient, the court should be petitioned to appoint a guardian to act on the patient's behalf.

4. In the event that there is no one to speak for the patient, then the patient's health care providers should assume that role.

Interdisciplinary teams need strategies for working with difficult families, especially when the client is incompetent (Molloy et al., 1991; Orlander, 1999). Here are some guidelines:

• Schedule a time for all family members to attend a team meeting. As many family members as possible should take part, as well as all professionals involved.

• Prepare a summary of the client status, in clear, consistent language. Do not use jargon or complex medical terminology.

• Describe the possible points of decision making with regard to treatment.

• State the dilemma clearly and ask the family to make the decision. The family must come to a consensus.

• Ask the family to designate a spokesperson to speak with the treatment team. Communicate any changes in client status or prognosis through the spokesperson.

- Give the family a deadline to make the decision.
- If the family cannot reach consensus, work with family members. If this is not effective, ask one member to begin to make the decision. Begin with the spouse if possible.

There are other situations in which case managers are working with families, especially when the client's skills are limited, but the client is not incompetent. Examples are clients with limited mental capacity, homeless individuals, and clients with mental illnesses. In working with each population, the case manager must maintain a delicate balance, advocating for the client, engaging the support of the family, and assessing what is in the client's interest. Many times families have invested years of effort and care in assisting the client. They may feel they deserve to have a primary role in the decision making (Fisk, Rowe, & Laub, 2000; Munro, 1997).

The following case illustrates some of the difficulties involved.

Reid was diagnosed as paranoid/schizophrenic at the age of 20. Now 30 years of age, he continues to struggle with his illness. His parents and younger sister have kept him at home with them for the past ten years, but living at home has been interrupted by periodic stays at the local state facility. Today Reid and his case manager have broached with the family the possibility of his living in an apartment in the city. The family's reaction is one of shock and anger. They refused even to discuss the issue.

The case manager can use some guiding principles to think through situations in which families and clients do not agree on treatment plans (McDonald, 2001). The case manager has a specific role in the decision making. He or she must ensure that the mission and goals of the organization are upheld, that professional ethics are considered, and that each member of the family is given the opportunity to be heard. In Reid's situation, the case manager works to establish clear communication in which the family can outline current conflicts and establish a plan supported by all parties.

 ## Working with Potentially Violent Clients

Cases of violence are increasingly common in the human service delivery system. The increasing incident of violence in the schools has amplified this concern. **Violence** is defined as "any physical or verbally assaultive behaviors, including hitting, biting, punching, choking, pinching, scratching, throwing, pushing, cursing, threatening, striking, or injuring with a weapon or item. . . . Violence is also physical harm to one's self or others, or the physical destruction or damaging of property" (Distasio, 1994, p. 14). Individuals who have the potential for violence are often human service clients. Since they have complicated and possibly long-term problems, their services are usually coordinated by a case manager. Distasio (1994, pp. 1–2) gives the following example of what can happen.

> *Harriet is a psychiatric social worker who was employed on the psychiatric unit of a large teaching hospital. She had been concerned about the anger her presence elicited from Mark, a 30-year-old chronic paranoid schizophrenic. He was very verbal in expressing his anger toward her. Mark had been granted weekend passes so that the staff could evaluate his ability to live in the community. After his last weekend pass, Harriet saw Mark standing in the hallway, conversing with another patient. Harriet continued walking down the hall, and as she neared him she caught a glimpse of something long and bright. In the next few seconds, Mark moved quickly, closed the distance between them, and stabbed Harriet in the abdomen. Harriet spent two weeks in intensive care, was on sick leave for the next two months, and needed extended psychotherapy to help her cope with the trauma of what occurred. . . . Harriet wonders if she will ever again be able to successfully confront what has become for her a personal demon—her fears and anxieties about working with violent or potentially violent patients.*

Case managers have a responsibility—to themselves and everyone else involved—to look for violent tendencies in their clients. Since it is the case manager who gathers much of the information about the client and monitors client treatment and progress, he or she is in a position to know of any history of violence or warning signs during treatment. The case manager then warns the interdisciplinary team of the possibility. If any other members of the team report danger signals, the case manager passes that information to all involved. It may also require changes in the treatment plan or the addition of other professionals to the team.

Our society is more violent today than ever before. Stresses are exacerbated by personal pressures, including divorce, illness, hopelessness, and lack of skills for coping with difficulties. In addition, the policies of deinstitutionalization and treatment within the least restrictive environment have released into the community many individuals who do not have resources to care for themselves. The problem is compounded by a lack of social support for many individuals, changing values, easy access to weapons, and a tendency in our culture to condone violence.

Some human service settings are more susceptible to violence than others. The most vulnerable seem to be psychiatric settings, nursing homes, emergency departments, and outpatient settings. In these settings, there is often insufficient support for staff who work with potentially violent clients.

Many agencies do not recognize the potential for violence in their clients. Service providers are not educated about this potential, and protocols for working with violent clients are seldom established. Many agencies just expect their clients to be violent. It is "part of their culture," and staff members are simply expected to deal with it. Some agencies reward staff for keeping things quiet, putting the emphasis on controlling clients rather than working with them in a therapeutic way. In such an environment, professionals may hesitate to report any signs of violence, fearing that they will be accused of not doing their jobs. Administrative factors may also contribute to poor management of violent clients: understaffing and insufficient supervision (Distasio, 1994).

The case manager can support the work of the team by acknowledging the potential for violence and preparing for manifestations of it. What exactly can the case manager do to address the issue of violence? The following eight steps, adapted from Distasio (1994), can be used as guidelines.

1. Help the members of the interdisciplinary team understand violence as a process: how to identify a preassaultive stage, what patient characteristics are predictors of violence, what helping situations increase the likelihood of violent behavior, and how staff behaviors and characteristics relate to violence.

2. Define aggression and the factors that are associated with it, such as lack of control, lack of freedom, and confusion.

3. Agree on intervention goals, such as calming the client, defusing the situation, and ensuring the safety of the client and others.

4. Teach intervention strategies to meet the goal of reducing violence or previolent behavior.

5. Provide guidelines for professional behavior.

6. Develop team responses to violent behavior.

7. Develop ways to document any violent events.

8. Encourage the agency to develop protocols for violent behavior. These may include a statement of institutional policy, a process of documentation, employee follow-up, and counseling for affected employees.

Confidentiality

In the helping professions, the obligation of **confidentiality** is fundamental to developing a relationship between the helper and the client. When the client is assured that information disclosed during the helping process will be kept in confidence, he or she feels free to share concerns and issues. The fuller the disclosure, the greater the opportunity for the case manager to gather valuable information about the client and his or her situation, which facilitates assessment and treatment planning. Trust between the helper and the client is a prerequisite to the success of their relationship. Case managers are in a unique position with respect to confidentiality, for they work with the family and friends of the client as well as with professional colleagues. As discussed in Chapter 4, this requires a special sensitivity to confidentiality issues.

One of the first points of discussion between case manager and client must be confidentiality and its meaning within the case management process. Four standards for confidentiality must be stated.

1. The case manager will keep the confidences of the client, except in instances of information about harm intended to self or to others.

2. The case manager will share basic information with colleagues on the team; he or she will inform the client who those colleagues are and what that information is.

3. With the consent of the client, the case manager will give family and friends certain information (Kane, 1999).

4. In most states, communication between case managers and clients is not legally privileged. That is, the case manager can be compelled to testify in court about what the client has said.

5. Confidentiality often includes issues of consent to request information and consent to release information.

Even with these guidelines articulated, dilemmas concerning confidentiality often arise. Consider the following situations.

An 18-year-old has just been diagnosed as pregnant. There is no legal obligation to inform the parents of the young woman or the father of the child, but the care coordinator always encourages clients to disclose this information. This young woman refuses. The discussion is closed and the matter remains confidential, even though the information is requested by young woman's mother.

An elderly man is furious with his social service coordinator because she told his daughter that he was dying of pancreatic cancer. The service provider knew that the daughter's husband had just asked for a divorce. The daughter was devastated because she had been counting on the father's support.

A counselor has been asked by his minister for information about a client who is a member of their congregation.

Having sensitive information about clients often causes dilemmas for case managers. To avoid problems, some decide to limit what information they seek to gather about a client: The less information they have, the fewer confidentiality problems they will encounter. Certainly, case managers must carefully choose what information to seek, but strict limits on information gathering are not always recommended. A complete history of the client enables the case manager to develop a treatment plan that will meet the needs of that client. *Relevant* information must be gathered, which helps the case manager understand how to work with family members and friends. Since the case manager is also coordinating care and monitoring progress, abundant information also helps him or her give better guidance to other professionals on the team.

The case manager should address certain matters before collecting any information, to anticipate any confidentiality problems (Kane, 1993, p. 150).

1. What does the case manager need to know to do the job, and why?
2. What should become part of the permanent record?
3. What understanding should exist between case manager and client about why the information is being sought, how it will be used, and the client's right to refuse to answer?
4. Under what circumstances should information be shared?

At the beginning of the process, the case manager determines the information needed to determine eligibility, conducts a comprehensive assessment, sets priorities, and develops a treatment plan. During the course of any case, information will emerge that the case manager needs to know but would not have thought to ask about at the outset. A good outline of the information to be gathered keeps him or her focused on appropriate and relevant areas to probe. The client often asks the case manager why certain information is necessary; at other times, he or she discloses much more than is needed for the process to proceed.

Sometimes the case manager needs information about the client from other agencies or institutions. It is important for the case manager to obtain written consent from the client to receive this information. The client may also have to provide a written consent to release the information to the agency or institution holding the information. To obtain client consent, the case manager discusses the need for the information with the client and helps the client with the appropriate paperwork. At this time, it is appropriate for the case manager to talk with the client about security of records, including those acquired from another agency.

Another dilemma is how much information to put in the permanent record. The case manager must record all information that documents the work with the client and describes his or her history. Not everything the client reveals needs to be recorded, however, especially if it is not relevant to the presenting problem. Unfortunately, the confidentiality of written information is not always secure, regardless of whether the record is on paper or computerized.

Once the case manager has decided what information to gather and what to record, he or she explains why each piece of information is important. Then he or she decides what parts of the information will be shared with family, friends, and other professionals and discusses this with the client. However, professionals and family members are bound to ask questions beyond what the client and case manager have agreed on. The client and the case manager negotiate about each such piece of information, and the client must give permission for disclosure. He or she also has a right to refuse to answer any question; if the information is necessary for determining eligibility, the case manager explains this. He or she can ask the client whether there is another way to obtain the necessary information.

Even following these guidelines, the case manager is bound to face issues that warrant further consideration. Three examples are the short cases presented earlier. Let's consider how these situations might be resolved. In the first case, the care coordinator gathers confidential information about the 18-year-old's pregnancy. The information remains confidential unless the agency has a policy mandating disclosure to parents or to the father of the child. If this is so, the coordinator should have informed the young woman at the time of intake. The care coordinator's obligation to inform supersedes the confidentiality guarantee. Such a policy is less likely to apply here since the young woman is of legal age. The situation becomes complicated if the young woman's mental competence is in question or if her health or that of the fetus is threatened in any way. If there is no legal or policy mandate, the coordinator must not inform the mother of the pregnant young woman, even though she is the potential grandmother.

In the second situation, the social service coordinator is torn between her allegiance to the dying father and her responsibility to his daughter. The difficulty here hinges on the coordinator's definition of who the client is. She has chosen to behave counter to the wishes of the father by breaking confidentiality with regard to his physical condition. Before doing so, she should ask herself the following questions: Is the father competent to request that his daughter not be told? Did he give the coordinator other information indicating that the daughter should be told, in spite of his reaction after the fact? Does she believe that it is in the father's best interest for the daughter to have this information? Does the coordinator see the daughter as the primary client? If so, why? The coordinator should not violate the father's request for confidentiality unless the answers to these questions provide sufficient justification.

In the third case, the counselor is asked for information by a professional who is not involved with the client's case, at least within the established service delivery system. The counselor is under no obligation to give the information unless there is an established need to know and the counselor gains the client's consent to share the information. It would be a different matter if the counselor had reason to believe that the client might harm himself or others. There would then arise a *duty to warn,* changing the counselor's obligation from confidentiality to a duty to share information. Before discussing a duty to warn in more detail, let's look at the client confidentiality issues that have developed in the last decade relative to technology.

Confidentiality and Technology

As case managers increase their use of technology to communicate, store, and retrieve information and care for clients, considerable concern is developing about client confidentiality. For example, many case managers today are using e-mail to communicate with other professional colleagues, to contact case management team members, and to explore referral services. By using the management databases established by human service agencies and organizations at the national, state, and local levels, case managers are able to enter and retrieve information from on-line databases. This means that while working on-line, case managers have access to comprehensive, up-to-date information about their clients. Case managers are also using organization and community computer networks to facilitate the sharing of information.

For many case managers, the computer has already become an integral part of their day-to-day work. As the use of technology continues to expand, case managers need guidelines to address the concern for and protection of client confidentiality. Several measures address security issues (Gahtan & Mann, 2001). Organizations can use systems with special encryption programs that scramble data and protect the data en route. They can install firewall systems for data stored on the Internet to prevent casual users and hackers from retrieving this information. To ensure the legitimacy of remote users, they can use special devices to authenticate the individuals who are using the system.

There are two other considerations about confidentiality for case managers to consider. The first focuses on the use of e-mail in service delivery. Because e-mail has become such a popular medium of communication, many patients are beginning to communicate with their case managers on-line. Sometimes clients are following up on a referral or on their treatment, just checking in, or requesting their own records. Indeed, in many instances, access to e-mail has increased the efficiency of communication between the case manager and the client. For clients who live in isolated, remote rural areas, live alone, or are homebound, e-mail and Internet allows a new type of access to services. If a case manager decides to use e-mail as a method of communication, some decisions must be made that address issues of confidentiality. Sherman and Adams (1999) suggest that helping professionals determine at the outset what types of case management services will be provided via e-mail. They need to obtain informed consent from the patient to communicate via e-mail, including a warning that, at times, e-mail may be sent to an incorrect address. They can state that once information leaves the case manager's office, its security cannot be guaranteed. Professionals might want to inform the client when a message is received and when the client might expect a reply. Finally, it is important to be professional in correspondence.

A second consideration also relates to technology and its influence on patient confidentiality. Technology enables case managers to work at home or at a remote site. Using fax machines, telephone answering machines, and computers with Internet access, case managers can write reports, make referrals, monitor client progress, and talk with clients while offsite. Using the home or a nonoffice site requires sensitivity to client confidentiality, especially violations of confidentiality that are inadvertent and unintentional (Woody, 1999).

Whether at home or at a remote location, what makes confidentiality more difficult is that the work space is usually shared. For example, when a case manager uses computers or Internet accounts that are also used by other family members or professionals, there is a risk to client confidentiality. Handwritten notes, reports, and files may also be left in this shared space. Care must be taken by the case manager to store these papers, either by securing a file or drawer or by taking the records to a secure place. Several safeguards for a shared work space include compiling a list of all technology you use, evaluating the risks of each to patient confidentiality, making changes in the procedures that contribute to the risk, and discussing with those with whom space is shared ways to respect the confidentiality of one another's work.

Duty to Warn

A case manager works in a nursing home for the elderly. One client, Mrs. Eddy, constantly expresses anger toward her husband. He lives at home and comes to visit her twice a day. He appears devoted to her and is her only contact with the world outside the nursing home. Today Mrs. Eddy says that she has a gun and plans to shoot her husband. The case manager is unsure whether to believe her. All the rooms are cleaned three times a week, and it is doubtful that a gun would remain unnoticed. Also, Mrs. Eddy is bedridden and has severe arthritis in her hands and fingers.

The **duty to warn** arises when a helping professional must violate the confidentiality that has been promised a client in order to warn others that the client is "a threat to self or to others" (Costa & Altekruse, 1994). The *Tarasoff* court decisions in 1974 and 1976 established that mental health practitioners have a duty to warn not only anyone who works with the client and the police, but also the intended victim. The *Tarasoff* rulings govern the law of most states (Fulero, 1988). It is a difficult judgment for a case manager to break confidentiality and invoke the decision to warn because confidentiality is such a fundamental responsibility. Violation of confidentiality is otherwise considered unacceptable practice in most instances.

The law and professional codes of ethics provide guidance in this matter. Further court rulings have clarified the duty to warn foreseeable victims, as first prescribed in the *Tarasoff* decision. However, state laws on professional confidentiality may conflict with the duty to warn (Costa & Altekruse, 1994). Also, professional codes of ethics do not always provide clear guidelines in this matter. Since case managers do not have their own professional code, they use the code of whatever discipline they hold credentials for. For instance, the *Code of Professional Ethics for Rehabilitation Counselors* states:

> Rehabilitation counselors will take reasonable personal action, or inform responsible authorities, or inform those persons at risk, when the conditions or actions of clients indicate that there is clear and imminent danger to clients or others after advising clients that this must be done. Consultation with other professionals may be used where appropriate. The assumption of responsibility for clients must be taken only after careful deliberation, and clients must be involved in the resumption of responsibility as quickly as possible. (Woodside & McClam, 2002, p. 346)

Members of the helping professions are encouraged to use codes as initial guidelines, but ethical codes are not written to cover every possible situation. They are established as principles, which professionals should use to guide their behavior. Such principles sometimes conflict and often do not address new issues that emerge.

The duty to warn is especially difficult for the case manager. He or she must consider the legal implications in case someone is hurt. In addition, there are complications in dealing with this issue in the context of the team.

Since the case manager sometimes provides direct services and also coordinates the treatment plan, the following questions may arise.

When do I give information that is related to duty-to-warn issues to other professionals who work with the client?

Do I tell individual professionals on the team on a need-to-know basis, or do I provide the information to the entire team?

What standards do these other professionals have concerning the duty to warn and confidentiality?

How do I involve the client in determining the answers to the preceding questions?

What guidelines does my agency have for confidentiality and duty to warn? Do these conflict with my professional code of ethics?

What are the guidelines of the other professional organizations represented on the interdisciplinary team? Are these in agreement with those of the professional organizations I belong to?

What legal standards are applied to the duty to warn and confidentiality in my state?

Because this issue is so complicated, Costa and Altekruse (1994) make the following suggestions.

Get informed consent. Many case managers make it standard practice to discuss the concept of confidentiality with clients at the outset. Because of the legal ramifications, many case managers now ask the client to sign an agreement stating that the case manager will report any information suggesting that the client may harm self or others (Margolin, 1982).

Plan ahead using consultation. The case manager must thoroughly research before working with clients and discuss it with the supervisor and other agency officials familiar with relevant policy and procedure (Corey, Corey, & Callanan, 1998). The issue should also be discussed with the team to raise everyone's level of awareness and develop approaches for addressing these situations.

Develop contingency plans. The case manager must think through exactly how he or she will handle a client who threatens harm to another and who claims to have a weapon available (in the home, office, or car).

Obtain professional liability insurance. Many helping professionals either have their own professional liability insurance or are covered through the agency's liability policy. It is important to know if the agency has such a policy; consult with supervisors about this.

Involve the client. There are advantages and disadvantages to involving the client if the case manager believes that the situation warrants a duty to warn. One way to involve the client is to remind him or her of your duty to violate confidentiality when there is a threat to harm self or others. The case manager can ask the threatening client to warn the intended victim. Although involving the client in this way can serve to maintain the trust of the relationship, it may also enrage the client and put the helping professional in danger. The client may feel cornered and become even more determined to carry out the destructive act.

Obtain a detailed history. The case manager's development of a detailed client history should continue throughout the process. As information is acquired, special attention should be paid to any indications of past acts of violence and evidence of impulsivity or anger. Look for other clues or warnings of danger, such as threats, a plan, or a weapon.

Document in writing. The case manager should document any indications of violence, threats, or expressed anger. Such notes are to be written as objectively as possible.

Costa and Altekruse (1994, p. 349) give guidelines for case managers when dealing with clients who represent a threat to themselves or others.

- Remind the client of your ethical and legal obligation to warn.
- Invite the client to participate in the process if possible.
- If possible, develop a plan with the client to surrender any weapons.
- Inform your supervisor, your attorney, law enforcement personnel, the local psychiatric hospital, and the intended victims.
- Keep the client committed, or have him or her committed for treatment.
- Keep careful records of all actions taken.

Autonomy

Client autonomy is a fundamental value in the case management process. As discussed in Chapter 1, case managers must be committed to the principles of client empowerment and participation, as well as to the ultimate goal of the process—that clients become able to manage themselves. Clients are to be involved in the process as much as possible, and treatment plans must support client choice and promote self-sufficiency. At the same time, the case manager has the obligation to provide the client with quality services and to act in his or her best interests. Sometimes, the case manager may believe that what the client prefers is not in his or her interest; this situation can entail an ethical conflict.

In an effort to encourage client participation, case managers can regard **autonomy** in a broader sense of increasing the range of choices a client can make. *Positive autonomy* means that the case manager works to broaden and strengthen the client's autonomy. If client preferences will result in danger to the client or others, the case manager must find a way to make those preferences more appropriate.

There are several instances in which client autonomy is not an absolute priority; for example, when client preferences interfere with other clients or other helping professionals, when the client is not competent to make decisions, and when clients need protection from their own decisions. In such situations, the case manager must have a clear conception of the reasons that autonomy should be restricted. The following case exemplifies the issue.

Mrs. Zeno is married and has three children. She has been a client of Child and Family Services for the past two years. She is now going to school, studying for an associate degree. She intends to work full-time after she completes her education. According to Mrs. Zeno, her husband takes all the money that she receives from Child and Family Services and from welfare. He gives her a small allowance, from which she buys groceries and pays for the rent and utilities, but there is never enough money for clothes for her children or educational supplies for herself. Mrs. Zeno is reluctant to challenge her husband. She does not want to disrupt her home, and she needs her husband's support if she is to finish school. In her last meeting with her case manager, Mrs. Zeno firmly stated that she would do nothing about her husband's behavior. The case manager believes that Mrs. Zeno should, at the very least, talk with her husband about the problem.

In dealing with issues of autonomy and self-determination, asking the following questions can help the case manager and guard the client's autonomy.

What are the facts of the case? Before any autonomy issues can be thoroughly understood, the case manager must gather relevant information.

Can you understand the history of the case and the current situation, and view it from the client's perspective? It is important to understand the client's perspective, which determines his or her preferences and choices.

Is the client competent? It is difficult to determine competence. Clients may be able to make good decisions at certain times but not at other times. Also, they may demonstrate competence in one area of decision making but not in other areas. Only with great caution should a client be declared incompetent.

What does the client want? Learn exactly what clients desire. Inform them fully of all the alternatives available, and help them understand the consequences (positive and negative) of each alternative.

What are the barriers to what the client wishes? For many reasons, clients may not be able to get what they want. Sometimes, policies and regulations restrict the case manager's ability to support client choices. Often there are resource problems: The services are not available, or funds are insufficient to pay for them.

Can these obstacles be overcome? It is the responsibility of the case manager to break down barriers to client autonomy. When making any decision that violates client autonomy, the case manager must inform him or her of the reasons for making that decision.

What are the risks involved in allowing client autonomy? In determining whether to violate client autonomy, one must calculate the risks to the client and others. If a client's decision would cause harm to self or others, the case manager must overrule him or her.

What is the advice of the treatment team? The opinions of the treatment team must be sought and considered; these professionals have much at stake in their own work with the client. If any of the professionals involved do not regard client autonomy as a high priority, this issue must be addressed early in the case management process.

How can the case manager guard the values of the client? The values of the client should guide the planning of services. If the client has conflicting values, the case manager works with him or her to sort through the priorities.

Client Preferences: One Component of Autonomy

To develop the care plan, case managers take into account client preferences. In every part of the plan, client preferences receive priority, and everyone involved must respect those preferences. Sometimes this creates problems, as illustrated in the following examples.

In developing a plan for Ms. Toomey, the case manager has limited funds, but part of the money available is designated for health and recreation activities, three days a week. Ms. Toomey hates any mention of good health; she especially dislikes exercise and fresh air. Ms. Toomey really loves to go to the movies; she does not watch them at home. The case manager is trying to decide whether she should use the resources available to pay for an attendant to take Ms. Toomey to the movies three days a week (which is Ms. Toomey's preference) or to have the attendant walk with her in the park at least once a week. The case manager is not sure that she can justify movies three times a week to the funding source.

Mr. Krutch needs child care for his baby while he works during the evening at a factory. The available child care workers are almost all Latino and speak little English, but they work well with children. Mr. Krutch refuses to have a Latino person in his home or caring for his child.

The case manager should consider the following guidelines when faced with issues involving client preferences.

Ask what the client needs. The case manager must have a clear idea of the client's needs. The client's perspective is important in identifying these needs.

Ask why the client has a particular need. For each of the needs listed, articulate clearly why the client needs the proposed services or support. The client and the case manager then understand why each of the needs is important. During this process, the cultural background of the client must be considered (Williams, 1993). Too often, case managers do not realize the preferences of their clients, particularly when they come from different cultures.

Can each of the needs be addressed? It is possible that not all the client's needs can be addressed. Sometimes resources are limited, but the client usually has some choices available. The resources available may determine which needs can be met. The case manager must tell the client what is available, and he or she must also promote client interests by advocating the introduction of services that are not currently available.

How are client preferences met? Respect for client autonomy means that the case manager works to meet the preferences of the client. Creative advocacy and problem solving can help in finding ways of meeting client needs. By advocating institutional adjustments to client preferences, the case manager builds trust and strengthens the relationship. Within a good relationship, clients are more willing to make their preferences known.

When is a preference not legitimate? The case manager has a responsibility to follow the agency's policies—one client is not entitled to an undue share of the case manager's time or of any other resource at the agency's disposal.

Let's look at how to apply these guidelines to the cases presented earlier. With Ms. Toomey, the case manager faces a dilemma. Ms. Toomey does not want to take part in any activities involving exercise or good health. She has made it clear that she only wants to go to the movies. The funding agency insists that its money not be used to pay for three movies a week. The agency will only fund one movie per week; its criteria for funding eligibility require that there be a variety of activities and that they be related to physical health. In this case, the treatment team agrees with the funding agency. It suggests that an exercise program and an activity that will challenge her intellect will better support the goals set for Ms. Toomey. With this information, the case manager can discuss options with Ms. Toomey. The case manager is confident that she and Ms. Toomey will find a solution. She also knows that she has limited time to spend on the matter because she is responsible for 35 other cases.

The service provider working with Mr. Krutch to find child care is in a difficult position. When Mr. Krutch expressed his disdain for individuals of Latino origin, the care coordinator listened carefully. Two issues emerged. First, most of the available child care workers are Latino. They are very able and have excellent recommendations. Second, the agency has a carefully stated commitment not to discriminate in hiring based on race, gender, religion, or national origin. On the other hand, the care coordinator working with Mr. Krutch realizes that it is important for him to have a good relationship and to trust whoever cares for his child. In this case, the care coordinator decides to try to find a non-Latino child care worker, but she also gives Mr. Krutch a realistic picture of what candidates are likely to be available. She explores Mr. Krutch's reasons for his reluctance to hire a Latino worker, and she prepares him for the possibility that he may have to give such a worker a try. One criterion she used when deciding to seek a non-Latino worker was to ask what harm this might do to the client. Mr. Krutch was vehemently against such a hire, and going against his wishes would not have been helpful to the goals of the care process.

Autonomy and End-of-Life Issues

Earlier in this chapter we discussed the ethical issues of decision making that related to patients who are dying and incompetent, unable to make their own decisions about their own end-of-life care. Today case managers are becoming increasingly involved in the care of competent individuals at the end of their lives as they work with elderly in long-term care facilities, with terminally ill inmates in correctional facilities, with AIDS patients, with people who have chronic illnesses, and with children and families caring for those who are dying. To give their clients more autonomy regarding their end-of-life decisions, case managers are helping clients consider establishing advance directives that will guide end-of-life care.

According to the Partnership for Caring (2001), an **advance directive** "is a general term that refers to your oral and written instructions about your future medical care, in the event that you become unable to speak for yourself. Each state regulates the use of advance directives differently. There are two types of advance directives: a living will and a medical power of attorney." The **living will** often defines what type of medical treatment the client wants at the end of life and what medical intervention the client does not want used, such as ventilator, heart/lung machine, nourishment and hydration, and resuscitation when sustaining life is futile. The client's right to accept or refuse medical treatment is protected by law (Partnership for Caring, 2001). A **medical power of attorney** is a document that records the name of the individual designated to decide about medical care if the individual is unable to do so.

It is the case manager's role to explain to the client advance directives or to refer the client to someone who can. Case managers help clients consider end-of-life issues and discuss preferences with members of their families. In the absence of family members, the case manager can help the client identify a friend or professional to make these final decisions if the client is unable to do so. This is a form of advocacy and empowerment, helping clients ensure that their wishes are known. The client has the final choice about whether or not to have an advance directive and to communicate what those directives might be.

Aging with Dignity (2000) has developed a format, *Five Wishes*, for case managers and family members to use to guide the decision making with regard to **end-of-life care**. The Five Wishes program centers the discussion on five requests.

1. The person I want to make care decisions for me when I can't
2. The kind of medical treatment I want or don't want

3. How comfortable I want to be
4. How I want people to treat me
5. What I want my loved ones to know

Helping clients think about and make their end-of-life decisions before their final days supports client autonomy and helps avoid the ethical dilemmas that do occur when these decisions have not previously been made.

Breaking the Rules

One of the most difficult dilemmas that a case manager can face is whether to comply with laws, regulations, and rules of practice when these do not appear to meet client needs. Two roles of the case manager collide—representing the agency versus serving as the client's advocate.

Ms. Dimatto is a case manager at an AIDS center in a small community. One of her clients, Mr. Sams, is dying. Mr. Sams is living with his partner, and his family resides in a nearby town. He has a good deal of family support. Mr. Sams is bedridden now, and his partner works full-time and cares for him at night. His family shares the responsibility for his care during the day. His mother stays with him three days a week, and his father three days a week. There is one day when Mr. Sams is by himself. During the other days, his parents read to their son, talk with him, give medications, and feed and bathe him. Mr. Sams' partner takes part in a self-help partners' group sponsored by the AIDS center. The center also works with the family, helping with any crises that arise and finding needed resources. In addition, the agency gives the parents a small stipend for the care that they give their son during the day.

The family is facing a problem. Next month, the mother is having bypass surgery, and she will need her husband to help her during her recovery. Neither she nor her husband will be available to provide care to their son. Mr. Sams knows that he needs daily care, so he has consented to hire an aide to stay with him and give him the help he needs. According to agency policy, however, his parents cannot receive any agency support for any care they give him for six months after the aide is hired.

To protect the client, the agency has established a rule that members of the family who receive in-kind or monetary support for client care must be providing such care on a continuous basis. If the care is interrupted for more than 14 days, the family cannot be supported by the agency as caregivers for six months. This rule works for stability of care and discourages families from providing help only when it is convenient for them. Another policy states, "The assistance must be provided by an individual whose sole responsibility is the care of the client."

Thus, in this case, the father may not be reimbursed if he cares for both his wife after her surgery and for his son.

Ms. Dimatto, the case worker, has a conflict to resolve. On the one hand, in accordance with the rules, she could hire an attendant to stay with Mr. Sams until his death. His parents could then visit him as they wish, but they would not receive any compensation for the care they give their son, nor could the father receive a stipend for caring for his wife. Ms. Dimatto could also suggest that the parents return after 14 days, asking the father to care for both his wife and his son. This solution would violate another policy—that the care of Mr. Sams be the father's sole responsibility—since the father would also be taking care of the mother during the postoperative period.

This case exemplifies the tension between the case manager's responsibilities as an employee of an agency and as an advocate for client interests. One question to ask (Kane, 1999) is, "Is it right or responsible to break rules for the welfare of the client?" Regulations established by governments and agencies are made to apply to everyone and every situation, but the reality is that few rules cover all situations. The obligation to follow regulations must be considered on a case-by-case basis. Kane (1993) proposes some guidelines for such consideration.

> *What is the purpose of the rule, and how does it help the client?* It is assumed that the rules have been developed to keep the client from harm. In Mr. Sams' case, the rule that encourages continuity of family care and penalizes any break in that service is written to discourage families from supporting the client only when it is convenient. The rule supports the importance of stability in care, especially for those who are ill and dying. The regulation that requires the attendant to have this client's care as his or her sole responsibility is likewise formulated for good reason—to ensure that the client receives maximum attention.
>
> *Who made the rule? What are the consequences of violating it?* Rules originate from many sources: federal, state, and local governments; local associations; and individual agencies. Each carries authority and imposes consequences for violating its rules. The case manager must know the source of the rule and the consequences of violating it. In Mr. Sams' case, the rule was instituted by the agency two years ago because families were taking agency stipends but were not providing good care for their relatives, who were the agency's clients. The agency finally decided that it was more important that clients receive quality care than that the care be given by family members.
>
> *What does the client lose and what does the client gain if the rule is followed?* The case manager carefully articulates the client's gains and losses in the event that a rule is broken. The weighing of advantages and disadvantages helps the case manager think through the issue and place client welfare at the center of consideration.
>
> *Is this a life-or-death situation for the client?* The answer to this question often gives perspective to consideration of the rule's effect on client welfare. If

the answer to this question is yes, the case manager has an argument worth considering for bending or breaking the rule.

Asking for help. In breaking or ignoring rules for the client's benefit, the case manager must keep in mind three considerations: good decision making, use of supervision, and asking for exceptions. He or she must ask the questions posed above to have a firm basis for decision. When using supervision or asking for an exception, violation of the rules may be only one of several concerns. When asked for help, the supervisor may respond in one of several ways. First, he or she may tell the case manager not to violate the rule. At that point the case manager must decide whether to obey the rule and the supervisor or to break the rule and ignore instructions; there are negative consequences for both actions.

Second, the supervisor may give the case manager permission to violate the rule. If the case manager does so, both the case manager and the supervisor are liable for the consequences. The supervisor may also ask the case manager not to act until an appeal for an exception has been filed.

Appealing for an exception. Appealing for an exception to a rule is a very different action from just bending a rule or violating a policy. The issue becomes more public. There is open dialogue about the rule and its purpose, the possible precedent of granting an exception, and the actual issues of the case for which the exception is requested. An appeal involves a foray into the realm of political activity; it is advocacy at a different level.

Petitioning for an exception can provide an impetus for change. It helps those who make the rules see that rigid enforcement may violate the reason for which the rule was made in the first place. In some situations, there is no way exceptions can be made. It can be useful to appeal for an exception even when there is no clear channel for such an appeal. Authorities may then understand the need to establish appeal procedures since rules are not appropriate in every situation that arises.

 ## Chapter Summary

Numerous ethical and legal issues surround the case management process. The issues arise within a context that makes decision making complex. Factors such as the increasing involvement of family members in client care and the reduction of resources available to assist clients increase the number of ethical issues case managers confront.

Family disagreements are one set of issues that case managers often face. Since families and friends are involved in client care, they expect to have a voice in the planning and treatment process. Frequently, there are strong disagreements among families. The "daughter from California" syndrome is one type of disagreement that occurs when the client is not competent to make decisions and the siblings disagree about treatment. When working with clients who are competent, it is also possible to have family disagreements. Case managers can develop the sensitivity to anticipate when these situations might arise.

Another issue relevant to the process of case management is working with potentially violent clients. Violence is much more likely to occur in some settings, and there are ways that individuals, agencies, and organizations can address the prevention of violence. Agreeing on intervention goals once the client has become violent, teaching case managers intervention skills to defuse violent situations, and developing team responses to violence are all good strategies to address a very difficult issue.

Confidentiality presents ethical dilemmas for case managers regardless of settings and client populations. Standards of confidentiality must be upheld, such as keeping confidences unless the information concerns harm to self or others. Some major decisions relative to confidentiality that the case manager makes are what should be included in a permanent record, what information should be shared with professionals in other agencies, and what information should be discussed with family and friends of the client. Technology has also made the keeping of confidences more difficult, but strategies exist to make client information more secure.

The duty to warn is an exception to keeping client confidences. This concept, clarified by the *Tarasoff* court case, states that if the case manager knows that the client is intending harm to self or others, the professional must alert appropriate individuals. Many situations have no clear-cut answers, and the case manager must make a judgment about the true intention of the client.

Autonomy represents a basic value of case management and a goal of the case management process. Case managers are committed to the principles of client empowerment and participation. We present a series of questions that guide the case manager when determining how autonomous the client can be. "Is the client competent?" "What does the client want?" "What are the barriers to what the client wishes?" A special case of autonomy that is becoming increasingly important is end-of-life decision making. Case managers are working with more clients who are competent but are dying and feel the need to determine their fate in their last days.

A case manager often has dual responsibilities to the client and to the agency. Codes of ethics, the law, agency policies and procedures, and professional values and judgment are not all in agreement. The case manager must determine what the rule or guideline for practice is, and in some circumstances, whether or not it should be broken.

Chapter Review

◆ *Key Terms* ....

Violence
"Daughter from California" syndrome
Duty to warn
Autonomy

Advance directive
End-of-life care
Living will
Medical power of attorney

◆ Reviewing the Chapter ···

1. What pressures make it difficult for case managers to confront ethical issues?
2. Describe the main issue in the "daughter from California" syndrome.
3. List ways that case managers can include family members in the case management process.
4. Why is violence so prevalent in human services today?
5. List eight steps case managers can follow to confront potential violence.
6. Explain the importance of confidentiality in the case management process.
7. How does confidentiality affect recordkeeping?
8. Describe how technology has affected confidentiality.
9. How does the duty to warn relate to confidentiality?
10. Describe how the case manager can experience conflict between the duty to warn and the requirement of confidentiality.
11. List the considerations for the case manager when giving duty-to-warn information to the interdisciplinary team.
12. Why can it be difficult to grant client preferences?
13. Relate the case of Ms. Toomey to the guidelines for thinking about issues of autonomy.
14. Why is a discussion of end-of-life issues important?
15. How does the case manager determine when it is okay to break the rules?

◆ Questions for Discussion ·······································

1. Why do ethical issues arise from the involvement of families in the case management process?
2. Confidentiality involves some special difficulties. Discuss them. What conclusions can you draw?
3. Do you think that it is a good idea always to grant client preferences? When is it not appropriate?
4. What advice would you give a new case manager about breaking a rule to meet a client's need?

References

Aging with Dignity. (2000). *Five Wishes*. Tallahassee, FL: Author.

American Counseling Association. (2001). *ACA code of ethics and standards of practice*. www.counseling.org/resources/codeofethics.htm

Corey, G., Corey, M. S., & Callanan, P. (1998). *Issues and ethics in the helping professions*. Pacific Grove, CA: Brooks/Cole.

Costa, L., & Altekruse, M. (1994). Duty to warn guidelines for mental health counselors. *Journal of Counseling and Development, 72*(4), 346–350.

Distasio, C. A. (1994). Violence in health care: Institutional strategies to cope with the phenomenon. *Health Care Supervision, 12*(4), 1–34.

Fisk, D., Rowe, M., & Laub, D. (2000). Homeless persons with mental illness and their families: Emerging issues from clinical work. *Families in Society, 81*(4), 351–359.

Fulero, S. M. (1988). Tarasoff: Ten years later. *Professional Psychology, 19,* 84–90.

Gahtan, A., & Mann, J. F. (2001). Technology and patient confidentiality. www.gahtan .com/alan/articles/techpat.htm

Kane, R. A. (1999). Choosing an adult foster home or a nursing home: Residents' perceptions about decision making and control. *Social Work, 44*(6), 571–585.

Kane, R. A. (1993). Uses and abuses of confidentiality. In R. A. Kane & A. L. Caplan (Eds.), *Ethical conflicts in the management of home care: The case manager's dilemma* (pp. 147–158). New York: Springer.

Margolin, G. (1982). Ethical and legal considerations in marriage and family therapy. *American Psychologist, 7,* 788–801.

McDonald, M. (2001). A framework for ethical decision-making. www.ethics.ubc.ca/mcdonald/decisions.html

Molloy, D. W., Clarnette, R. M., Braun, E. A., Eisemann, M. R., & Sneiderman, B. (1991). Decision making in the incompetent elderly: The "daughter from California" syndrome. *Journal of the American Geriatrics Society, 39*(4), 396–399.

Munro, J. (1997). Using unconditionally constructive mediation to resolve family system disputes related to persons with disabilities. *Families in Society, 78*(6), 609–616.

Orlander, J. D. (1999). Use of advance directives by health care workers and their families. *Southern Medical Journal, 92*(5), 481–484.

Partnership for Caring. (2001). Advance directives: Living wills, durable power of attorney for health care. www.choices.org/ad.htm

Sherman, L., & Adams, M. (1999). Patients and e-mail: Technology means increased technology concerns. *Wisconsin Medical Journal.* www.wismed.org/wmj/may_june99/mayjne99_sherman.htm

Williams, O. J. (1993). When is being equal unfair? In R. A. Kane & A. L. Caplan (Eds.), *Ethical conflicts in the management of home care: The case manager's dilemma* (pp. 206–208). New York: Springer.

Woodside, M., & McClam, T. (2002). *An introduction to human services* (4th ed.). Pacific Grove, CA: Brooks/Cole.

Woody, R. H. (1999). Domestic violations of confidentiality. *Professional Psychology, Research and Practice, 330*(6), 607–610.

Surviving as a Case Manager

As a case manager, how much time do I have to deal with this family? How much bonding am I going to do? If I know that I have a year, how much am I going to disclose and how close am I going to get?
> —Zulma Resto, caseworker, Casita Maria Settlement House, Bronx, NY, personal communication, May 4, 1994

We have individuals living here from Iran and Iraq. They don't deal with our culture at all. They have their own doctors, dentists, and they have their own way of dealing with "crazy" people. They stick together. They are the shopkeepers and people who run the small businesses. They encapsulate and they bring people from the other country. They save up amongst themselves, and they have a job open, and they groom this person to come over. The Vietnamese also have their own way of doing things. And what looks like child abuse to us was not child abuse to them. We got all sorts of referrals from schoolteachers about these kids. Parents beat kids with sticks on the back to drive out evil demons, not to abuse the child.
> —Walter Rae, social worker, Houston, TX, personal communication, December 16, 1999

I got on my soap box about certain things. People don't like it. They do not want to hear about it. Welfare changed and I think I was right that there would be more homeless people on the streets. . . . We have got to start handling that situation.
> —Anais Watsky, Northwest Assistance Ministries, Houston, TX, personal communication, December 16, 1999

The passage of federal legislation, the development and growth of the managed care industry, and the evolution of the human service professional have contributed to the reemergence of case management in human service delivery. You read in Chapter 1 about the reasons for its reemergence. One reason is a response to the shift in service delivery from large, state-operated institutions to community-based services. Coupled with this shift is the resulting dual nature of service delivery that requires both service provision and service coordination in order to meet client needs. Finally, the managed care industry, in its efforts to control the quality and cost of health care, has had a significant impact on terminology. Its use of the terms *case management* and *case manager* has increased their visibility.

You have also read about the evolution of case management as a service delivery strategy that reflects these changes. Some aspects of case management that are different today from 20 years ago are the identity of the case manager, the necessary knowledge and skills of those who deliver services, the types of problems clients experience, and the nature of the bureaucracy. Case management as a service delivery strategy will continue to evolve, making use of technology, recognizing new client problems, and responding to the changing economic and political climate.

The purpose of this chapter is to integrate what you have learned reading and studying this text with the wisdom of case managers across the United States who are engaged in service delivery. They offer real-life lessons, as Zulma Resto does in her chapter-opening quote, about surviving as a case manager today. Her words illustrate the continuing struggle with boundaries, time, and caring. Case managers are constantly facing challenges as they work with diverse populations. Walter Rae illustrates the difficulties of providing services to diverse cultures. He suggests that services be tailored to fit the unique characteristics of people from another culture. Working for a nonprofit organization that uses volunteers to provide help in the community, Anais Watsky takes a leadership role as she attempts to persuade the community to address current and future social service issues.

We think you will find that the knowledge and experiences of these case managers affirm the case management concepts introduced and discussed in this text, suggest new areas of emphasis, and identify new skills. To help you understand the application of the information shared here, the chapter includes a case study that is a composite of experiences of several case managers. For each section of the chapter, you should be able to accomplish the following objectives.

THEMES IN CASE MANAGEMENT TODAY
- List the themes that case managers identify when they talk about their professional work.
- Define what case managers mean by a Jack or Jill of all trades.
- Illustrate why communication is important in case management.
- Describe the use of decision-making and critical thinking skills in case management.
- List three personal qualities that help case managers do their jobs.

CASE STUDY
- Describe five interactions that take place in the case concerning the guardianship of baby Juan.

SURVIVAL SKILLS
- Describe five characteristics of burnout.
- List ways that burnout can be identified and prevented.
- Explain why time management is an important skill for the case manager.
- Describe four steps in establishing a time-management system.
- Define assertiveness and explain why it is important.
- Illustrate ways in which the case manager can be assertive.

Themes in Case Management Today

Case managers whose voices you will "hear" as you read this chapter represent social service agencies across the United States. They identify themselves as having the job title of case manager or they describe their primary job responsibility as case management. They may have different job titles, such as caseworker, social worker, family advisor, or behavior specialist. Our analysis of the interviews we conducted with 50 to 60 human service professionals resulted in the articulation of eight common themes. These themes, describe what they do and what they need to know how to do to be effective case managers. The themes respond to the complex and sometimes difficult situations case managers encounter as they cope with large caseloads, clients with multiple needs, and scarce resources. In addition, case managers often work with clients who are silent, reluctant, or resistant in a bureaucracy that requires detailed documentation for each interaction. The eight themes are (a) the performance of multiple roles, (b) organizational abilities, (c) communication skills, (d) setting-specific knowledge, (e) ethical decision making, (f) boundaries, (g) critical thinking, and (h) personal qualities.

Performance of Multiple Roles

Every day, as case managers work with their clients, they perform multiple roles, including advocate, broker, coordinator, educator, evaluator or assessor, gatekeeper, monitor, planner, and problem solver. (You read about these roles in Chapter 3.) It helps to be a **Jack or Jill of all trades**. Many helpers combine the evaluator/assessor, planner, broker, and coordinator/monitor roles into an intensive case management function and assume responsibility for determining the real issues, developing care plans, finding resources, and coordinating care among other professionals. Problem solving is often part of case management and occurs when everything is going smoothly as well as when there is a crisis. Problem solving requires a plan A, a plan B—and a plan C, if necessary. For many helpers, the final goal of managing cases is self-sufficiency or the resolution of issues for the client.

Advocacy, discussed in Chapter 8, is an example of the complexity of roles in case management. One interviewee suggested that "services begin with advocacy," and another described the advocacy role as a means to "instill in clients what is best for them." Advocacy occurs when case managers are fighting for quality services for their clients, helping families treat their members fairly, working with agencies and other bureaucracies to serve clients better, and supporting clients "when they can no longer even support themselves." One helper's approach to advocacy was "to teach clients how to deal with their problems . . . how to deal with the system." On another level, interviewees defined advocacy for the agency as fighting for resources, attracting clients, and representing the agency.

The successful case manager is the professional who has multiple skills and is able to use them as needed, sometimes simultaneously. The complexity of a particular role coupled with other job demands often makes this especially challenging.

Organizational Abilities

Several case managers emphasized the importance of **organizational skills**, and they mentioned the disasters that can occur when professionals are not organized.

> Theoretically speaking, if you have a case load of 30 families and each one of those families has an average at any given moment of 10 things that they need you to work on, then that is 300 needs that you are juggling . . . any of which could explode.

They are also aware that if they are not organized, it is their clients who will suffer. For them, being organized means managing time and completing paperwork.

The concept of time influences case management in a variety of ways: organizing, budgeting, scheduling, responding, balancing, and slowing down. Case managers recognize the need for time to "let things percolate up," " sit down and remember everything you have done during the week with that client," and "take one step at a time." Even though time management was consistently mentioned as a tool to alleviate stresses and pressures, many admitted that the workload is so "horrendous" that they never gain control over their work situations. They say that they are so busy, they don't have time to seek a resolution to a client's problem. One participant said, "Clients have to come to me." The difficulty with time management arises from several sources, including the unpredictability of the workday, external deadlines, and the ever-changing bureaucracy.

Completing paperwork or documentation is another organizational skill that case managers practice daily. They understand the purpose of the paperwork that flows through their offices and its importance in building records, making requests, accounting for expenditures, documenting a "client's whole life on paper," jogging the memory, and providing an audit trail. The documentation they describe includes initial assessments, family histories, psychosocial assessments, contact notes, goals, service plans, and evaluations. Many of these are discussed in this text. The extensiveness of paperwork is illustrated by the following quote.

> We keep a copy of everyone's birth certificate, Social Security card, Medicare card, Medicaid card . . . IHP plan from their initial staff meeting . . . yearly and disciplinary plan . . . if person has achieved a goal . . . even to cutting their toenails, they have to have a medical form.

These professionals face the dual challenge of knowing how to write reports and how to find time to do their paperwork. In addition, they struggle to "set up a work space where you can find . . . 5,000 pieces of paper needed at any moment."

Organizational skills, then, pervade each day's work providing case management services. Even though many helping professionals choose this field because of the client contact, they are often surprised at the responsibilities of case managers, including the required documentation. If no organizational skills are mastered, the overload, frustration, and eventual burnout forces many to leave the profession. Organizing time, paperwork, caseloads, daily schedules, and emergencies is a critical survival skill.

Communication Skills

The case managers we interviewed claimed that "communication is more important than any other skill" and is tied directly to establishing a helping relationship, assessing needs and situations, and selling and persuading clients. Central to these results are the **communication skills** of listening, questioning, and persuading.

These case managers talked about "really listening" to what is being said by the client in order to establish a helping relationship, identify problems, and move through the case management process. For them, bonding is an important part of the helping process because the case manager must get clients' "trust . . . so they will tell you what the real problems are." Without this relationship, clients are less likely to accept services or to continue to reach out over a long period. Case managers form strong attachments to their clients; several agencies support clients through death, and case managers can be involved in arranging and attending funerals and "sharing grief with the family."

Good listening helps these case managers understand and evaluate what the client is saying (e.g., "Is the client telling you about the real problem?"). Listening also makes a therapeutic contribution to the client's progress: "Often times a client will start talking to you and . . . will ask . . . questions and answer the questions in one breath." Listening helps some clients let off steam. In other instances, clients have multiple problems that are jumbled together. "You can listen so you can see questions that will help them be more clear about what they mean."

Questioning is another communication skill that facilitates the assessment process. It involves "gathering information from lots of people," "assessing people's needs," and "figuring out what is important and what is not." Interviewees discussed the art of questioning, emphasizing asking "the right questions" to gather information for social and client histories, needs assessments, and intakes. Questions enable case managers to "look in every corner," assess the situation as well as client needs, and make appropriate decisions regarding case management. (Questioning was discussed in Chapter 5.)

To persuade clients to become involved in receiving services and to be active in their own self-care is an important communication tool. Sometimes case managers have to "sell the mother on letting her daughter be independent" and "sell parents whose children are involved in the criminal justice system on helping the case manager."

Clients today are consumers of services, since many of them, under the auspices of managed care and other funding models, are able to choose where they will receive the services to which they are entitled. In other words, "it goes back to client choices . . . it is the consumer's choice to participate." Therefore, some agencies find themselves competing with other agencies for clients, so they spend time convincing clients to come to them for services.

Communication with other professionals is also viewed as a valuable skill. Networking is an important tool for finding resources for clients. Having a good relationship with other professionals and knowing which person or agency to call for help ultimately benefits clients. A second part of communication with other professionals involves relating to other staff, especially other team members. For most of these case managers, working with their colleagues is "pretty family oriented, everybody is real close." These professionals use each other for support so they don't feel "they are just hanging out there" alone. They are also working on problems that demand "getting a lot of heads together" to solve problems. For many case managers, the following quote summarizes the feelings they have about the people with whom they work: "I have survived these past two and one-half years . . . [by] establishing a relationship with my team."

Setting-Specific Knowledge

Case managers also mentioned setting-specific knowledge as critical to their job performance. This includes general skills, such as typing and computer usage, as well as more specialized knowledge (e.g., medical terminology, medications and their side effects, drug regimens). For others, a thorough understanding of human behavior, psychosocial issues, and various helping methods form a basis from which to work with clients, make assessments and recommendations, and develop plans. Also important is knowing how systems, such as Medicaid, probation, child welfare, and housing, help the case manager support the client's interaction with other agencies and services. Finally, many helpers believe that "street smarts" are essential for case management. Street sense allows case managers to provide realistic assistance to clients very different from themselves.

Ethical Decision Making

Another case management skill discussed by the case managers who participated in this study was **ethical decision making** (discussed in Chapter 10). They must be able to identify ethical issues (e.g., self-determination, confidentiality, and role conflict), to ask the questions that surround the issues, and to make appropriate professional responses. One case manager summarized the issue of self-determination this way: "People have the right to poor judgment. . . . we can educate them, but we can't take away their rights." The case managers interviewed gave many examples of incidents in which, in their opinion, clients made bad choices. Clients refused services, chose to retain independence rather than receive complete services, refused medications, ate poorly, returned to abusive situations, and violated parole agreements. Case managers often described their frustration when clients refused services or did not heed sound advice, but they were passionate about the rights that clients have to determine their own destinies.

One ethical dilemma concerns confidentiality. For many of the case managers we interviewed, there is often a question of what information goes into a report and what should be omitted. Another issue of confidentiality emerges as computers are more widely used. One case manager expressed a hope that computer security is receiving the attention that confidentiality demands. This concern was addressed in Chapter 10.

One of the most difficult dilemmas that case managers face is role conflict. They describe it as "working both sides of the fence," "walking a thin line," or "protecting two sides." When role conflict occurs, case managers are asked to assume dual responsibilities that may be at variance with one another. One helper shared a role conflict she encountered as a parole officer. The parole officer's primary responsibility is to support the parolee's life outside the prison environment. If the parolee "messes up," the parole officer becomes the prosecutor. This type of conflict makes it difficult to maintain a supportive, trusting relationship with a parolee after the prosecution.

Boundaries

Surviving the intensity of helping was a concern to the case managers in this study. They felt it was essential for the case manager to establish **boundaries** between self and client. One helper explained, "You have to watch yourself because you get too attached to clients." Another elaborated: "You have to have an idea of your own boundaries and what your issues are. You don't want to get yourself confused about what is going on. . . . Sometimes it's helpful to step back and ask, 'Whose problem is this?'" Even though case managers work hard to establish boundaries between themselves and their clients, they still agonize over their clients and the difficult situations they face. One interviewee described it as "close detachment." One case manager described her reactions to client problems thus: "Sometimes I have had to go to the restroom to cry for a few minutes because it was a really hard case." Many workers even dream about their clients. These professionals are committed to handling boundary issues and recommend "staying realistic" and "leaving work at work."

Critical Thinking

The effective case manager is one who needs to think critically and clearly. One of the **critical thinking** skills needed is "seeing the whole in addition to individual, narrow parts." Because individual cases are so complex, there is a danger of focusing on the details and "not seeing the forest for the trees." For one case manager, critical thinking is "being able to procure and digest a large body of information from all kinds of different people" and then determining the real issue. The existence of underlying issues means that case managers are "detectives" who are able to ask the right questions, not take things exactly as they are presented, conduct continual assessments, and assess communication as it is happening. Helpers then "put it all together to get a better picture." Interviewees suggested that using their years of experience, tuning into their insight, "going in with a clear mind," and "performing a reality check—my fantasy versus reality" all enhance their critical thinking abilities.

Personal Qualities

The people we interviewed identified a number of personal qualities that enable a case manger to be effective, including *realistic, patient, flexible*, and *self-confident*. Two interviewees likened the desirable personal qualities to those of a fairy godmother and a chameleon.

Flexibility was a consistent recommendation. Constant interruptions to the plans for the day, unpredictable and emergency client needs, interviews conducted under unusual circumstances, and interactions with different clients and professionals who demand different styles are examples that illustrate the need for flexibility. Case managers also need to be "firm" in communications and "soft" at other times. One participant explained, "I just kind of roll with the punches. Whatever needs to be done, I just do it."

The ability to form good working relationships with clients is essential. For some case managers this is carried to an extreme of "somehow being reasonable to the point where you can no longer be reasonable." Others describe it as being tactful and respectful of others, "going out of the way to communicate with others," and "taking time with people." The case manager is in a people-oriented field, and whether working with clients or other professionals, it is important to "be able to talk with people and get along with them."

Another necessary quality for case managers is patience. One participant described her willingness to "take one step at a time" when working with her clients. According to these case managers, it is sometimes difficult to be patient. They reported that they remind themselves and their colleagues of the importance of "being able to let go and wait." Part of patience comes from being realistic, including having realistic expectations for clients. These helpers also realized that "when you are allowing clients to learn to help themselves, you cannot be in a hurry." A second factor in patience is persistence. For many professionals, "you just keep plugging." They acknowledge the difficulties and the resistance encountered from both clients and bureaucracies during the helping process.

Self-esteem provides the foundation for many of the difficulties encountered in case management. According to participants, "you must have self-confidence." It helps to maintain a positive perspective when things do not go well: "You realize that you will get over it [failures] and things will move on." This confidence also fosters **assertiveness** when dealing with other professionals or resistant clients. One special challenge of working with other professionals is the bureaucratic hierarchy; many higher-ups do not acknowledge that case managers have important professional contributions to make. Self-esteem also helps them assume a leadership role when it is required and, as you read in Chapters 8 and 10, leadership is required more often today. These case managers recognized the need to assume authority in order to accomplish their goals.

Finally, according to participants, case managers need to have both a "sense of adventure" and "excitement" about their work. Terms like *anthropologist, private eye*, and *detective* described the challenges of case management: Human service providers must accurately identify problems, develop service plans to meet client needs, provide and seek out services, and evaluate the process. For many, the challenges are stimulating, not depressing; these people thrive on working in a demanding, fast-paced environment.

 ## Case Study

The following case study illustrates many of the themes just discussed. It is based on several cases that occurred in the southeastern United States. This case crosses state lines, involves a number of professionals, and illustrates the realities of providing services to an ethnically diverse population. The case is complex as well as confusing, reflecting the reality of case management. Since it is told from the point of view of the case manager, Delores Fuentes, her narrative is in italics. As you read the case study, identify situations that illustrate the themes described in the previous section.

People Involved in the Case

Delores Fuentes	Case Manager, community health and human services clinic
Mr. and Mrs. Ruiz	Parents of Juan
Dr. Hidalgo	Pediatrician
Ms. Brown	Nurse, Health and Rehabilitation Services, Georgia
Mrs. Marcos	Guardian in Minneapolis
Mexican Consulate	In Atlanta
Mr. Sanchez	Lawyer, Mexican Consulate, Georgia
Dr. Stapleton	Physician in Minneapolis
Mexican Consulate	Minneapolis
Police—first encounter	Minneapolis
Mr. Gluckey	Staff member, Mexican Consulate, Minneapolis
Police—second encounter	Minneapolis

My name is Delores Fuentes, and I am a case manager at a community health and human services clinic in rural Georgia. Since I work with a diverse group of clients, including migrant farm workers, my work is very challenging. One problem we face is the client's lack of understanding of the culture here, of the standards of living in this country, and of the laws that govern relationships, such as marriage and child custody. Let me share a recent case in which there was such a misunderstanding.

Mr. and Mrs. Ruiz came to the clinic to see the pediatrician, Dr. Hidalgo. Mr. Ruiz asked the doctor if he was willing to take care of his baby, Juan.

DR. HIDALGO:	I know you want me to see Juan. Where is he?
MR. RUIZ:	Oh, I need you to sign these papers saying that you are willing to take care of him so they can release him from the hospital in Minneapolis.
DR. HIDALGO:	Okay, but why is he there?
MR. RUIZ:	Oh, he was sick.
DR. HIDALGO:	So he is there in the hospital?
MR. RUIZ:	No, he is with a translator. The nurse from Health and Rehabilitation Services (HRS) needs to have a paper from you because if not, we will lose our baby.
DR. HIDALGO:	I don't understand this. Do you mind if I send you back to HRS?

Dr. Hidalgo asked me to look into this case so he could understand the situation better. Altogether it took me about two hours to figure out what was going on. The scrap of paper Mr. Ruiz gave the doctor had a phone number and a name on it—that was it. From there I started trying to figure out the case. First, I called the HRS in the state capital and said, "I have this family here and I need to speak to the director of nursing." Eventually, I spoke with a nurse, Ms. Brown.

| MS. BROWN: | Don't even bother to take on the case. It is a lost cause. It was referred to me by a hospital in Minneapolis [where Mr. and Mrs. Ruiz had recently been living]. They wanted to make sure that the family has a suitable home for the baby. I made a visit; they live in a house with no electricity and no phone. There are three or four people living with them, so that is not a healthy environment for the kid. So I signed that Juan cannot be released into their care. |

I asked her Juan's location now. She reported that he had already been released and is with a guardian. I needed the name of the guardian, and finally she gave it to me. I reported to the pediatrician, and he told me that he would call the guardian. He reported this conversation.

| DR. HIDALGO: | My name is Dr. Hidalgo and I am a pediatrician working at a community health clinic. I need some information about Juan Ruiz. Could you tell me what is going on with Juan? |
| MRS. MARCOS: | Don't even bother asking questions of the parents. They don't have any more right to be trying to fight for him. I have custody of Juan. |

I wondered how that could be. How is that possible? She should have been able to describe why she had custody. She could have said, "I have custody because the father mistreated the wife, and I was able to remove that kid from that unhealthy environment."

I called Mrs. Marcos back and told her that the pediatrician had spoken to me. I asked her what the issues are, and she told me not to assess the situation. She explained that she had observed the interaction between Mr. and Mrs. Ruiz. According to her, Mr. Ruiz is the only one who talks—his wife always looks down, she cannot face you. Mrs. Marcos concluded that it was a typical case of abuse.

In Mexican culture, it is a mark of respect for the lady not to look her in the eyes, and it is the man who answers the questions. I told the pediatrician what I had learned and that I would visit Mr. and Mrs. Ruiz to determine whether it was an abusive situation. I talked with them and their neighbors and friends and found no evidence of abuse.

I then called the Mexican Consulate for help, explaining that one of our clients had a problem and describing the details I had uncovered so far. I said that I thought the family needed a lawyer. They agreed. After investigating the case, the lawyer, Mr. Sanchez, called me.

MR. SANCHEZ: I found out that the couple signed custody papers for a Mrs. Marcos to take care of the baby, Juan, while he was in the hospital. Nothing was signed after that. The hospital released the baby without permission from the parents to this person who actually, at least initially, had the good will to help. She was a translator and spoke Spanish well, so the family thought she would help them. I think initially that was what she really intended. Later she became attached to Juan.

I called Mrs. Marcos back. I explained that the case is complicated and that although I do not have any legal background, I had found a lawyer for the family. I told her that she would receive a call from a lawyer working for the Mexican Consulate. He was exploring the issues and wanted her to tell him more about the situation. She said that Mr. Sanchez had already called her. I think she was becoming nervous because she called Dr. Hidalgo. She told him she wanted to take care of the baby and that Juan needs this and that. Put like that, it sounded like the baby was almost dead. Dr. Hidalgo asked me to contact the Minneapolis hospital for the name of the doctor. I talked with the doctor, Dr. Stapleton, who confirmed that nothing was wrong with the baby when he was released. I gave his name and phone number to the lawyer, Mr. Sanchez. Dr. Stapleton told Mr. Sanchez what he told me: "There is nothing wrong with the baby." Then Mr. Sanchez told me about his conversation with Mrs. Marcos.

MR. SANCHEZ (to Mrs. Marcos):	The doctor said that the baby must go home to Georgia.
MRS. MARCOS:	Oh, no. I will not release the baby.

Eventually, an agreement was reached that the baby, the guardian, the family, and the lawyer would all see Dr. Stapleton. Meanwhile, Mr. and Mrs. Ruiz learned that Mrs. Marcos had been seen putting some boxes in her car, and they called me at home at 2:00 A.M. I called Mr. Sanchez, who said that we needed to send the couple to Minneapolis immediately. In the meantime, he would call the police.

We sent the parents to Minneapolis in a van. While they were on the way, we made sure the family had a house in Georgia to return to, as well as work, a pediatrician for the baby, and a phone. They arrived in Minneapolis the next morning. The family reported that the translator, Mrs. Marcos, was very nice.

MR. RUIZ: She give us a piece of cake and a coffee, and let us play with the baby. Then she disappeared. About ten minutes later the police came in, and the police literally chased us out, and said, "If you come back here, we will put you in jail."

Mr. Ruiz called me from a gas station to report what had happened, and I told him to call me back in an hour. I called the Mexican Consulate to report on the situation. The Consulate here called the one in Minneapolis and asked them to have someone meet the family at the gas station. We called the police to tell them what was going on. The police agreed to check on it.

The police sergeant found a report with the correct time and date, stating that the parents were removed from the home of Mrs. Marcos, but he did not have the full report of what happened. So a member of the Mexican Consulate in Minneapolis and their lawyer, Mr. Gluckey, went to see Mrs. Marcos to get custody of Juan. All agreed to meet at the doctor's office the next morning, and if he said the baby could travel, then Mr. and Mrs. Ruiz could take Juan back to Georgia.

This case took three weeks to resolve.

 ## Survival Skills

The Ruiz case illustrates the complexities of case management. It involves a set of parents (in Georgia), a pediatrician and a case manager (Georgia), a nurse (in the state capital), a translator turned guardian (Minneapolis), a doctor (Minneapolis), two Mexican consulates and their lawyers (Georgia and Minneapolis), two teams of police, and an indeterminant number of people involved in facilitating and obstructing the case management process. Many individuals were helpful; some were passive or indifferent. Several individuals did not trust the clients, the parents of the child. Some professionals believed that the child was the client, not the parents.

The activity in this case reflects many of the themes described earlier in this chapter, such as the performance of multiple roles, the ability to communicate, knowledge of ethical decision making, and the ability to be flexible and patient. The case also suggests how difficult performing the role of case manager can be. It emphasizes the stressful and emotional nature of the work that can sometime leads to burnout. Ms. Fuentes, the case manager, saw her work come to a positive end. It is one of her best successes. Often she works hundreds of hours with little reward for her efforts.

The skills of time management and assertiveness can provide the case manager with stability and support to counter many of the challenges of this difficult work. These are survival skills. The following section discusses the nature of burnout and describes how time management and assertiveness can prevent or alleviate it.

The Prevention of Burnout

Burnout was first defined by Christine Maslach (1978) as a phenomenon experienced by those who work in the human service area. Current research on burnout underscores Maslach's findings (Woodside & McClam, 2002). Those most vulnerable to burnout are "people whose work in one way or another involves direct contact with different kinds of recipients—welfare clients, patients, children, prisoners . . ." (Maslach, 1978, p. 56). According to Maslach, **burnout** is the "emotional exhaustion resulting from the stress of interpersonal contact" (p. 56). Professionals with case management responsibilities are particularly susceptible to burnout because many of the factors that contribute to burnout are integral to their work. Among those factors are the nature of the clients, the stresses of dealing with bureaucracy, and a personal tendency to react negatively in stressful situations.

Delores Fuentes, who worked so hard to reclaim Juan for his parents, experienced many of the work characteristics that can lead to burnout. In this case, she did not react negatively. On the contrary, she was able to pursue her advocacy for Mr. and Mrs. Ruiz beyond the bounds of her agency's responsibilities and across state lines.

Much of the work of case management involves clients who have very complex and long-term difficulties: children who are at risk, adults who have disabilities or are elderly, and people with medical problems, such as AIDS or cancer. In fact, modern-day case management emerged as a methodology designed to handle the overwhelming stress of serving clients with multiple problems.

Case managers daily come face to face with clients who have little hope and who often confront long-term illness and severe social difficulties. Other case managers work with clients who demand intense services over long periods. Often, no one asks if the case manager has the time or the resources to provide the services needed.

Delores Fuentes certainly experienced a situation that required her immediate and undivided attention. Her other work was left undone. In addition, she was working in "uncharted waters," with a variety of professionals from other states and other international governmental agencies.

Case managers also struggle with the bureaucracy in which they work. One pervasive difficulty is the size of caseloads. Case managers often lack sufficient time to give each client quality treatment; they must settle for services that are adequate at best. The programs offered and the resources available often do not match clients' needs. Case managers find themselves in the middle—understanding clients' needs all too well, yet knowing the limitations of the system.

As Delores Fuentes struggled to provide a fair hearing for the parents, she found that the bureaucracy was ill equipped to handle a situation that involved non-U.S. citizens across state lines. Ms. Fuentes had to forge new alliances with agencies such as the Mexican Consulate.

Certain observable symptoms related to work habits, psychological and physical well-being, and relationships with clients can indicate that case managers are reacting negatively to stress. Absenteeism, tardiness, and high job turnover rates are several indicators of widespread burnout. A case manager might also have symptoms such as gastrointestinal problems, substance abuse, exhaustion, sleep disturbance, poor self-concept, inability to concentrate, and difficulties in personal relationships. Their attitudes and working patterns may change: They become disorganized, have difficulty working with colleagues, blame clients for their own problems, and depersonalize clients by referring to them with disparaging names (Anderson, 2000; Maslach, 1978; Rohland, 2000; Woodside & McClam, 2002). These changes soon affect job performance, and case managers can feel increasingly frustrated and worn down by their responsibilities. Burnout is a devastating problem for an agency, the case managers affected, and the clients served.

Acquiring time-management and assertiveness skills can help case managers avoid burnout. Both skills allow them to be more effective case managers.

Managing Time

Time management is particularly important for case managers, but the nature of their work makes it especially difficult. Many types of activities are required of them: interviewing clients; writing reports; meetings with clients, team members, and others; arranging services; transporting clients; and monitoring services. Each of these mandates specific skills, a special orientation, and considerable time. In addition, the various activities require interaction with many different individuals, including colleagues, clients, and families. Each contact takes time—making the initial contact, monitoring the work, and maintaining relationships with all those involved. Another facet of the job that makes time management difficult is the uniqueness of each case. At the beginning of the work with a client, it is hard to predict the full scope of the services that will be needed or the intensity of support the client will need.

Another difficulty is the lack of *consistent* demands on time. As a result, it is very difficult to establish a regular schedule and to plan and then perform certain tasks at appointed times. When crises interrupt this complicated work, the case manager must constantly weigh and shift priorities. Even the best time manager must set aside the plan for the day to assume additional responsibilities that emerge and demand attention.

In thinking about managing time, Cassell and Mulkey (1985) suggest asking four questions:

- What personal characteristics, both strengths and weaknesses, influence how case managers manage time?
- Exactly how is time spent?
- What techniques are used to control time?
- What time-management system is used? What principles are followed?

Before case managers can manage time effectively, they must assess how they actually use time during the workday. A time record can be used to establish time spent and outcomes achieved. A time log records activity every 20 minutes or so, describes and summarizes interactions, and covers at least two weeks so as to capture variations in workflow. Once such a log is completed, the data are analyzed to reveal how time has been allocated.

There are several ways to analyze the data. One is to examine the distribution of time in various case management activities, such as intake interviewing, reading reports, team meetings, and making community visits (Cassell & Mulkey, 1985). It is also helpful to know how much time is spent managing cases and providing services. A form like the one in Figure 11.1 might be used for such an analysis.

Delores Fuentes spent three days working intensely for Mr. and Mrs. Ruiz and their baby, Juan. Because she managed a community health clinic and saw clients of her own, her other responsibilities had to be absorbed by her coworkers. At the same time, she had to be extremely organized to coordinate assessment, fact-finding, and the search for resources. Delores was the point person, receiving updates for all professionals working on this case. She made a plan each day, but since the events continued to change quickly, she was continually revising her plan.

	Number of Hours/Week	%
Managing		
Intake interviewing		
Report writing		
Team meetings		
Community visits		
Making referrals		
Monitoring the plan		
Community outreach		
Subtotal		
Direct service		
Direct service to clients		
Direct service to families		
Subtotal		
TOTAL		

Figure 11.1   Time allocated to case management activities (managing and direct service)

Time-Management Techniques

The time analysis of the tasks of one day, considered in light of the extent of control the case manager has, provides the basis for managing time. One component of time management is the planning and implementation of more effective use of time in several steps. It begins with setting goals and priorities and ends by assigning particular activities to times when they can best be completed.

Laekin (1976) recommends that desired goals be categorized into three time spans: the next five years, the next three months, and this day. The next step is to assign priorities to the goals. For example, an ABC scale can be used, on which A is the most important, B is moderately important, and C is least important. The goals can then be sorted into categories: short-term, intermediate, and long-range. How do case managers use this information in planning for the day? They focus on the most important activities and choose tasks that are ranked A (most important) from each of the categories—short-term, intermediate, and long-range.

The following guidelines also help determine how to plan each day.

THE RESPONSIBILITIES
- Pick the two most important goals and pursue them first.
- Address the toughest jobs first.
- Alternate difficult and easy tasks.
- Group similar tasks.

THE PLANNING OF TIME
- Plan some uninterrupted time.
- Plan time for the unexpected or for crises.
- Allot time according to deadlines.
- Assign time to plan.
- Assign time for paperwork.
- Have a list of quick tasks to use as filler.

TIME MANAGEMENT AND OTHERS
- Understand assignments.
- Delegate tasks that others can or should do.
- Develop a system for monitoring the work of others.
- Say no when the assignment is inappropriate or there is not enough time to complete the task.

Time management is one skill that can help the case manager establish priorities and assess the needs of a situation and the professional resources available. Assertiveness is another skill that can facilitate effective case management. Case manager Delores Fuentes used assertiveness skills as she worked with the parents, other professionals, and the guardian.

Acting Assertively

Communication is "a process of exchanging ideas and feelings. It occurs when meaning is conveyed from one person to another, verbally and nonverbally" (Hamilton, 1992, p. 338). Important communication between individuals occurs even when there is silence. Communicating effectively is always a challenge, and one's ability to communicate can always be improved. The work of case management can be facilitated by **assertiveness.** To communicate assertively is to express oneself with confidence and conviction (Lee & Crockett, 1994). Communicating assertively is also positively related to "building self-concept, increasing self-confidence, and decreasing stress and anxiety" (Lee & Crockett, 1994, p. 422).

There are several elements to communicating assertively. The first is recognizing the equal rights of the speaker and the listener. The second is maintaining a positive attitude about the communication and wanting it to be fair. An assertive communicator is also comfortable with negative feelings such as anxiety, anger, and frustration (Alberti & Emmons, 1986; Weinfeld & Donohue, 1989). The goal of assertive communication is to relate to others in ways that lead to maximum dialogue and a positive outcome for each participant. All who participate must know that they are respected and that their ideas have worth. This manner of communicating builds trust.

Assertive behavior can be better understood in contrast to two other ways of communicating: nonassertive (or passive) and aggressive. *Nonassertive communication* is characterized by inaction. The passive individual does not participate in the communication process but maintains silence or quietly acquiesces. In essence, such passivity negates any opportunity for dialogue that would help clarify, expand, or change the pattern of thought or action under discussion. Those who are passive often feel as if they have nothing of value to say or that they will be punished if they speak up. Passive communicators violate their own rights by allowing others to infringe on them (Alberti & Emmons, 1986; Lee & Crockett, 1994; Weinfeld & Donohue, 1989).

Delores Fuentes was assertive throughout this case. She assumed the role of advocate for Mr. and Mrs. Ruiz, tracking down information from Ms. Brown and pursuing the case when it was recommended that she drop it. She also contacted consulates and lawyers and engaged the help of the police. In fact, she did what she needed to do to help her clients. As she tells her story, it is clear that she understands the difference between passivity, assertiveness, and aggressiveness. She took action and she was assertive, but not aggressive, as she dealt with other professionals.

People who communicate aggressively do so in a way that violates the rights of other speakers and listeners. In contrast to assertive communication, aggressive communicating is hostile, competitive, and one-way. Aggressive communicators wish to dominate the interaction and to give orders; they are not seeking dialogue. They want their opinion to be the final word heard and to govern the decision. Such people stand up for their own rights in a way that violates the rights of others (Alberti & Emmons, 1986; Lee & Crockett, 1994; Weinfeld & Donohue, 1989).

Assertive communication works in the case manager's favor because every facet of case management involves communicating with others. Whether you are initiating assertive communication or responding to passive or aggressive clients or colleagues, assertive communication is an important skill.

Communicating assertively also has much to do with the principles of case management discussed in Chapter 1. Both empowerment and advocacy benefit from assertive communication. *Advocacy* means standing up for the rights of those who cannot represent themselves, and when case managers engage in advocacy, they are acting assertively.

Empowerment is the activity of instilling the belief that clients have rights and abilities and can solve problems for themselves. Assertiveness promotes the self-confidence and self-respect that supports empowerment. At times, clients and their families do not want to be empowered, which creates a very difficult situation for case managers. Clients may be angry when they do not receive what they feel they are entitled to or when they are asked to help themselves.

GUIDELINES FOR ASSERTIVE BEHAVIOR

Two guidelines are useful when learning assertive behavior: Respect personal feelings and know personal rights. When individuals are not acting assertively, one of these principles has usually been forgotten (Alberti & Emmons, 1986; Weinfeld & Donohue, 1989).

Respect personal feelings   When acting assertively, the communicator must respect his or her own feelings and those of others. For example, the case manager must make sure his or her view of a particular situation gets heard. People often just walk away from situations because they do not know how they are feeling, do not trust their feelings, or do not think that they have the right to those feelings. Exploring personal feelings, recognizing them as legitimate, and listening to others' feelings provide the basis for assertive communication.

Know personal rights   Each person has a number of rights that are the foundation of a sense of personal worth and dignity. Awareness of these rights and acceptance of them make it easier to learn how to be assertive.

Case managers can learn to be assertive and apply these skills in several ways. Sometimes case managers are asked by clients or colleagues to make judgments about particular matters, and sometimes they themselves feel the need to provide feedback. In both cases, there are ways of communicating that will be seen as helpful and thoughtful rather than critical or hurtful. Occasionally, the case manager must confront clients or other professionals. Assertiveness allows confrontation to occur in a way that respects the rights of both the speaker and the listener and promotes honest, fair, and positive outcomes.

It is often difficult for case managers to say no to requests from colleagues or clients. Their commitment to help others makes them inclined to say yes to any request. However, a request is unreasonable if the action would violate ethical or legal codes. Colleagues or clients may ask for action that causes case managers to be frustrated, anxious, or physically or emotionally threatened. To be reasonable, a request must take into consideration the rights of case managers, as well as those of clients and colleagues.

From time to time, everyone encounters people who communicate in an aggressive way. It is difficult to turn an aggressive encounter into an assertive one.

One's reaction may be passive because the aggressiveness is so surprising, and sometimes aggressiveness is returned for aggressiveness. Either type of response leaves the feeling that rights have been violated and that communication is closed. Acting assertively helps the case manager to turn aggressiveness into reasonableness.

Chapter Summary

This chapter provides a summary of eight themes derived from interviews with case managers, who talk about their work in human services. These reality-based themes illustrate the concepts presented in the previous chapters. These case managers emphasized the importance of performing multiple roles when working with their clients. For example, sometimes they are brokers, linking their clients to services, and other times they must advocate for new services that will help them meet their clients' needs. Other themes are survival skills, such as organization, communication, and critical thinking, which are important in effective case management practice. Organizational abilities allow case managers to handle large caseloads, work with clients who have multiple needs, maintain contacts and make referrals to various agencies and organizations, and handle the paperwork. Communication pervades the case management process and is essential in working with both clients and other professionals. Written communication, report writing, and case summaries all have their place in the case management process. Critical thinking skills allow the case manager to process information and make decisions by being clear, logical, thoughtful, attentive to the facts, and open to alternatives.

Ethical decision making and boundaries are directly related to the dilemmas and challenging situations that confront case managers each day. Both concerns require a recognition that problems exist and a willingness to work to resolve these problems. The personal qualities that are necessary to be effective in the job include being realistic, patient, flexible, and self-confident. The case manager must function in a variety of situations, under intense pressure, with very difficult and challenging clients.

In this chapter, Delores Fuentes, a case manager in a community health and human services clinic, presents the case of Juan. The case illustrates just how intense and complex the work of case management can be. It affirms and demonstrates the themes the case managers discussed in the interviews, emphasizing the complications that can occur when working with people from other cultures.

As Delores Fuentes' case demonstrates, case managers can have very stressful jobs. Too many clients, too few resources, unhelpful bureaucracies, and challenging clients are some of the factors that contribute to stress. Too much stress often results in burnout, which is accompanied by an inability to work with clients and others effectively. Two skills help counter burnout or prevent it. Time management is particularly important for case managers, but the nature of their work makes it difficult. Case managers can use guidelines and planning techniques to coordinate their priorities and goals in the time available.

Assertiveness is based on the principle of equal rights for speaker and listener. Often, the case manager must be assertive to procure services for a client and when working with a difficult client. The goal of assertive communication is to relate to others in ways that lead to maximum dialogue and a positive outcome for each participant.

Chapter Review

◆ *Key Terms* ....

Jack or Jill of all trades	*Critical thinking*
Organizational skills	*Assertiveness*
Communication skills	*Burnout*
Ethical decision making	*Time management*
Boundaries	*Empowerment*

◆ *Reviewing the Chapter* ....

1. Why are there times when case managers have to perform multiple roles?
2. What are the challenges in case management that require organizational abilities?
3. What are the ways that case managers can communicate with their clients and other professionals?
4. What are the ethical decisions that confront case managers?
5. Why are boundaries an issue for case managers?
6. Define critical thinking.
7. Describe the personal qualities that case managers need.
8. Describe the complexities of the case of Juan Ruiz.
9. Define burnout.
10. Why do case managers burn out?
11. What are the symptoms of burnout?
12. Describe how the time-management guidelines might have helped Delores Fuentes with Juan's situation.
13. Define assertiveness.
14. What are ways in which case managers may be assertive?
15. Describe how assertiveness principles might have helped Delores Fuentes in the case of Juan.

◆ *Questions for Discussion* ....

1. Describe five themes relevant to case management today. Provide an example for each theme.
2. Why is burnout an issue for case managers? Discuss how you would prevent burnout.

3. Are you a good time manager? Illustrate your self-assessment with three examples from your own life. What skills could you improve?
4. Why is assertiveness an important characteristic for the case manager? Describe a situation in which you would find it difficult to be assertive?

References

Alberti, R., & Emmons, M. L. (1986). *Your perfect right.* San Luis Obispo, CA: Impact.

Anderson, D. G. (2000). Coping strategies and burnout among veteran child protection workers. *Child Abuse and Neglect, 24*(6), 839–848.

Cassell, J., & Mulkey, W. (1985). *Rehabilitation caseload management.* Austin, TX: PRO-ED.

Hamilton, P. (1992). *Realities of contemporary nursing.* Reading, MA: Addison-Wesley Nursing/Benjamin-Cummings.

Laekin, A. (1976). *How to get control of your time and your life.* New York: Signet.

Lee, S., & Crockett, M. S. (1994). Effect of assertiveness training on levels of stress and assertiveness experienced by nurses in Taiwan, Republic of China. *Mental Health Nursing, 15*(4), 419–432.

Maslach, C. (1978). Job burnout: How people cope. *Public Welfare, 36,* 56–58.

Rohland, B. M. (2000). A survey of burnout among mental health center directors in a rural state. *Administration and Policy in Mental Health, 27*(4), 221–237.

Weinfeld, R. S., & Donohue, E. M. (1989). *Communicating like a manager.* Baltimore: Williams & Wilkins.

Woodside, M., & McClam, T. (2002). *Introduction to human services.* Pacific Grove, CA: Brooks/Cole/Wadsworth.

Glossary

Accountability Being responsible to another person for one's actions and use of resources.

Achievement test A test for evaluating an individual's present level of functioning.

Active listening "Hearing" both verbal and nonverbal messages, as well as what is not said—the thoughts and feelings of the person communicating.

Advance directive A general term for a person's instructions about future medical care in the event he or she is unable to speak for himself or herself.

Advocacy To speak or write in defense of a person or cause, pleading his or her case or standing up for his or her rights.

Advocate A person who speaks on behalf of clients when they are unable to speak for themselves or when they speak and no one listens.

Ageism Discrimination or expression of bias against someone based on the person's age.

Applicant An individual who requests services or who is referred for services of a human service agency.

Aptitude test A test that gives an indication of an individual's potential for learning or acquiring a skill.

Assertiveness Expressing oneself with confidence and conviction while respecting the rights of others.

Assessment A phase of case management that involves evaluating the need or request for services and determining eligibility for services.

Assessment interview An interaction that provides information for the evaluation of an applicant for services.

Attending behavior Ways in which a person communicates interest and attention; for instance, using eye contact, attentive body language, and vocal qualities.

Authority The ability to control resources and actions.

Autonomy The client's right to make choices.

Broker A role in which the case manager acts as a go-between, linking those who seek services and those who provide services.

Budget A numerical expression of an agency's expected income and planned expenditures for an upcoming period.

Burnout The emotional exhaustion resulting from the stress of interpersonal contact.

Case history interview   A comprehensive interview that includes open-ended questions and questions that require factual answers. It may include family history and a chronology of major life events.

Case management   A creative and collaborative method of service delivery, involving skills in assessment, consulting, teaching, modeling, and advocacy, that is intended to enhance the functioning of the client.

Case manager   A helping professional who provides direct services to clients, links clients to services, and monitors the process.

Case notes   A written account of each visit, contact, or interaction a case manager has with or about a client.

Case review   A periodic examination of a client's case.

Chain of command   The flow of authority in an agency or organization, from the position with the most authority to the position with the least authority.

Client   An applicant whose request for services has been approved by an agency.

Client empowerment   Developing the client's self-sufficiency to enable him or her to manage life without total dependence on the human service delivery system.

Collaboration   The process of working with others.

Collaborator   A role that involves relationships with other service providers to meet client needs.

Community organizer   A case management role focused on assessing community needs and planning ways to meet those needs.

Confidentiality   A guarantee to the client that information disclosed during the helping process will be kept in confidence.

Consultant   A professional who has expertise to support case management functions.

Content recording   A word-for-word account of what was said by each participant in an interview.

Continuity of care   Comprehensive care provided during and after service delivery.

Continuous improvement   A movement whose purpose is improving service delivery.

Contracts   Agreements that secure funding from governmental agencies or private corporations.

Coordinator   A role in which the case manager works with other professionals and staff to integrate services.

Cost containment   The process of planning resource allocation to prevent expenditures from increasing.

Counselor   Case manager role of therapist.

"Daughter from California" syndrome   Family disagreements over the care of an incompetent client.

Deinstitutionalization   Moving clients from self-contained settings to various community-based settings, such as half way houses.

Departmental team   A small number of professionals who have similar job responsibilities and support each other's work.

Documentation   Written presentation of data, observations, interviews, and services.

DSM-IV   The *Diagnostic and Statistical Manual* (4th edition), published by the American Psychological Association, which classifies all types of mental disorders.

Duty to warn   A situation in which a helping professional must violate the confidentiality that has been promised a client in order to warn others that the client is a threat to self or to others.

Effective communication   Verbal and nonverbal messages (through greeting, eye contact, and responses) that let the client "know" the interviewer.

Empowerment   The belief that all individuals, regardless of their needs, have integrity and worth. Because of this belief, the case manager places the client in a central role in case management.

End-of-life care   Attention to decisions concerning terminally ill family members or clients.

Equal access to services   Nondiscrimination in granting services. Because of the commitment to equal access, the case manager assumes the role of advocate and develops ways of extending services.

Ethical decision making   The ability to identify issues, such as self-determination and confidentiality, and make appropriate professional responses.

Ethnocentrism   The belief that one's own group has desirable characteristics and others outside the group are substandard.

Evaluator   One who collects information to determine client's functioning and to assess service provision.

Expediter   In this capacity, the case manager ensures that services are delivered in an efficient and effective manner.

Feedback logs   Records that provide feedback to the agencies that deliver services to help ensure quality information and referral services.

Fee-for-service   Payment for cost of service provided.

For-profit agency   An agency that provides a service and makes money.

Goal   A statement describing a broad intent or a desirable condition.

Halo effect   A favorable or unfavorable early impression that biases the judgment of the observer or interviewer.

HMO (health maintenance organization)   A managed care model that is very structured and controlled and emphasizes positive health promotion.

Implementation   The third phase of case management in which service delivery occurs; the case manager either provides the services or oversees their delivery.

Informal structure   The everyday way an organization does business, determined by personal relationships and influence and varying from traditional organizational structure and formal lines of communication.

Information and referral system   The knowledge of what services are locally available and how to gain access to those services.

Intake interview   A structured interview, usually occurring when a person applies for services, which is guided by a set of questions in the form of an application.

Intake summary   A type of summary recording that is written at some point during the assessment phase.

Integration of services   A coordinated effort on the part of many agencies and professions to bring together services to help a client.

Intelligence test   A test that measures intelligence in the form that the test creators define the term.

Interdisciplinary team   A team that includes professionals from various disciplines, each of which represents a service the client might receive.

Interview   A face-to-face meeting between the case manager and the applicant. It may have a number of purposes, including getting or giving information, therapy, resolution of a disagreement, or the consideration of a joint undertaking.

Job description   A general guideline that describes the work for which an employee is held responsible.

Living will   An individual's written instructions about medical treatment or intervention at the end of life.

Managed care   An agreement with providers of physical and mental health care to guarantee services.

Maximum performance test   A test for which examinees are asked to do their best at something.

Medical consultation   An appointment with a physician to interpret available medical data, determine any medical and vocational implications for health and employment, and recommend further medical care if needed.

Medical diagnosis   An appraisal of an individual's general health status to establish whether a physical or mental impairment is present.

Medical power of attorney   A document that records the name of an individual who can make decisions for another about medical care if this person is unable to do so.

Mental status examination   A structured interview consisting of questions designed to evaluate an individual's current mental status, considering factors such as appearance, behavior, and general intellectual process.

Mission statement   A summary of the guiding principles of an agency.

Mobilizer   One who works with other community members to obtain new resources or services for clients and communities.

Monitoring services   Reviewing the services the client receives, the conditions that may have changed since planning, and progress toward the goals and objectives of the plan.

Not-for-profit agency   An agency that does not make money.

Objective   An intended result of service provision.

Open inquiries   Questions that elicit broad answers, allowing the client to express thoughts, feelings, and ideas.

Organization-based case management   A model of case management that focuses on ways of configuring services that are comprehensive and meet the needs of clients who have multiple problems.

Organizational chart   A symbolic representation of authority, accountability, and information flow within an organization or agency.

Organizational climate The conditions of the work environment that affect how people experience their work.

Physical examination Inspection, palpation (feeling), percussion (sounding out), and auscultation (listening) by a physician. Typically, this is conducted from the skin inward through various orifices and from the top of the head to the toes.

Plan A document written before service delivery that sets forth goals, objectives, and activities.

Plan development A process that includes setting goals, writing objectives, and planning specific interventions.

Planner The role of the case manager when preparing for the service or treatment the client is to receive.

Planning The second phase of the case management, in which the case manager and the client develop a service plan.

POS (point of service) A managed care option in which clients are encouraged to use providers in the managed care system but do not lose all their benefits if they choose medical care outside the system.

PPO (preferred provider organization) A kind of health care plan that falls between the traditional HMO and the standard indemnity health insurance plan.

Pretreatment review The reviewing and approval by a managed care professional of a treatment plan before service delivery.

Privileged communication A legal concept referring to the right of clients not to have their communications with a professional used in court without their consent.

Problem solver The case manager role that assists in client-sufficiency, decision making, and problem solving.

Problem solving The process of identifying challenges and resolving difficulties.

Process recording A narrative account of an interaction with another individual.

Psychological evaluation An integral part of the client study process, having as its objective the understanding of the individual by measuring characteristics that pertain to behavior.

Psychological report A document that reports on the evaluation of behavioral characteristics and mental capacity.

Psychological test A device for measuring characteristics of human beings that pertain to behavior.

Public agency An agency that exists by public mandate and receives funding from federal, state, regional, county, or municipal governments.

Quality assurance Programs that focus on developing standards of care.

Quality care An emphasis on providing the best services to the client in terms of effectiveness and efficiency.

Racism Discrimination or unfavorable opinions based on an individual's race.

Record Any type of information related to a client's case, including history, observations, examinations, diagnosis, consultations, and financial and social information.

Referral Connecting the client to a resource within the agency or at another agency.

Reliability The degree of consistency with which a test measures whatever it is measuring.

Resource selection The process of choosing an individual, program, or agency to meet the client's needs.

Responsibility-based case management A model of service delivery in which case management can be conducted by the family, supportive care network, volunteers, or the client.

Role-based case management A model of case management that centers on the roles and responsibilities the case manager is expected to perform.

Role conflict The assumption of dual responsibilities that may be at variance with one another.

Second-opinion mandates The requirement that a second professional be consulted before a treatment plan can be approved.

Service coordination The process of locating, arranging, and supporting the client's use of community resources.

Sexism Discrimination or unfavorable opinions based on an individual's gender.

Social diagnosis A systematic way in which helping professionals gather information and study the nature of client problems.

Social history The telling of the client's story in his or her own words, with guidance from the helper, reflecting the client's life and individual characteristics.

Social service directory A catalog listing the problems handled and services delivered by other agencies.

Sources of error Potential biases that affect an interview's reliability and validity.

Staff notes Comments written at the time of each visit, contact, or interaction that any helping professional has with a client.

Structured clinical interview An interview consisting of specific questions asked in a designated order.

Structured interviews Direct and focused interviews, usually guided by a form or a set of questions that elicit specific information.

Summary recording An organized presentation of facts; the form of recording preferred by most human service agencies.

System modifier An administrative role, usually with authority to change agency policy or redirect priorities.

Teamwork Professionals sharing responsibility for clients.

Termination The final step in case management, when the client and the case manager review the problem, goals, plan, services, and outcomes.

Test A measurement device.

Treatment team A group of professionals who meet to review client problems, evaluate information, and make recommendations about priorities, goals, and expected outcomes.

Typical performance test A test such as an interest or personality test that gives an idea of what an examinee is like.

Unstructured interview An interaction that consists of a sequence of questions that follow from what has been said.

Validity The extent to which a test or interview measures what we wish to measure.

Verbal following Minimal verbal responses that let the client know you are listening.

Violence Any physical or verbally assaultive behaviors, including harm to oneself or others or the physical destruction or damaging of property.

Whole person A holistic view of the individual.

Word root The main part or stem of a word.

Index

TO THE OWNER OF THIS BOOK:

We hope that you have found *Generalist Case Management: A Method of Human Service Delivery* useful. So that this book can be improved in a future edition, would you take the time to complete this sheet and return it? Thank you.

School and address: _____

Department: _____

Instructor's name: _____

1. What I like most about this book is:_____

_____

_____

2. What I like least about this book is: _____

_____

_____

3. My general reaction to this book is: _____

_____

4. The name of the course in which I used this book is: _____

_____

5. Were all of the chapters of the book assigned for you to read?_____

   If not, which ones weren't? _____

6. In the space below, or on a separate sheet of paper, please write specific suggestions for improving this book and anything else you'd care to share about your experience in using the book.

_____

_____

_____

_____

_____

Optional:

Your name: _____ Date: _____

May Brooks/Cole quote you, either in promotion for *Generalist Case Management: A Method of Human Service Delivery* or in future publishing ventures?

Yes: _____ No: _____

Sincerely,

Marianne Woodside
Tricia McClam

IN-BOOK SURVEY

At Brooks/Cole, we are excited about creating new types of learning materials that are interactive, three-dimensional, and fun to use. To guide us in our publishing/development process, we hope that you'll take just a few moments to fill out the survey below. Your answers can help us make decisions that will allow us to produce a wide variety of videos, CD-ROMs, and Internet-based learning systems to com-plement standard textbooks. If you're interested in working with us as a student Beta-tester, be sure to fill in your name, telephone number, and address. We look forward to hearing from you!

In addition to books, which of the following learning tools do you currently use in your counseling/human services/social work courses?

_____ **Video** _____ in class _____ school library _____ own VCR

_____ **CD-ROM** _____ in class _____ in lab _____ own computer

_____ **Macintosh disks** _____ in class _____ in lab _____ own computer

_____ **Windows disks** _____ in class _____ in lab _____ own computer

_____ **Internet** _____ in class _____ in lab _____ own computer

How often do you access the Internet? _____

My own home computer is a:

_____

The computer I use in class for counseling/human services/social work courses is a:

_____

If you are NOT currently using multimedia materials in your counseling/human services/social work courses, but can see ways that video, CD-ROM, Internet, or other technologies could enhance your learning, please comment below:

_____

_____

_____

Other comments (optional): _____

_____

Name_____ Telephone_____

Address _____

School _____

Professor/Course_____

You can fax this form to us at (650) 592-9081; or detach, fold, secure, and mail.

FOLD HERE

BUSINESS REPLY MAIL
FIRST CLASS   PERMIT NO. 358   PACIFIC GROVE, CA

POSTAGE WILL BE PAID BY ADDRESSEE

ATT: *Marketing*

**The Wadsworth Group
10 Davis Drive
Belmont, CA 94002**

FOLD HERE

Attention Professors:

Brooks/Cole is dedicated to publishing quality publications for education in the social work, counseling, and human services fields. If you are interested in learning more about our publications, please fill in your name and address and request our latest catalogue, using this prepaid mailer. Please choose one of the following:

☐ social work   ☐ counseling   ☐ human services

Name: _____

Street Address: _____

City, State, and Zip: _____